T0122432

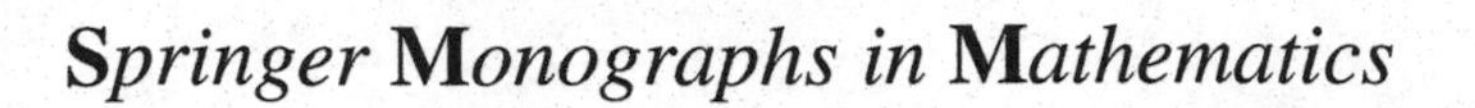

For further volumes:
http://www.springer.com/series/3733

Michael A. Henning • Anders Yeo

Total Domination in Graphs

Michael A. Henning
Department of Mathematics
University of Johannesburg
South Africa

Anders Yeo
Department of Mathematics
University of Johannesburg
South Africa

ISSN 1439-7382
ISBN 978-1-4899-9156-0 ISBN 978-1-4614-6525-6 (eBook)
DOI 10.1007/978-1-4614-6525-6
Springer New York Heidelberg Dordrecht London

Mathematics Subject Classification (2010): 05C69, 05C65

Printed on acid-free paper

Springer is part of Springer Science+Business Media (www.springer.com)

We dedicate this book to our wives, Anne and Angela, and to our parents, Joy and Cosmo Henning and Gunnel and Geoffrey Yeo, with heartfelt thanks, for their love, support, encouragement, and patience.

Preface

In 1998 Teresa Haynes, Stephen Hedetniemi, and Peter Slater [85, 86] wrote the two so-called "domination books" and in so doing provided the first comprehensive treatment of theoretical, algorithmic, and application aspects of domination in graphs. They did an outstanding job of unifying results scattered through some 1,200 domination papers at that time. The graph theory community is indeed much indebted to them.

Some 14 years have passed since these two domination books appeared in print and several hundred domination papers have subsequently been published. While these two books covered a variety of domination-type parameters and frameworks for domination, this book focuses on our favorite domination parameter, namely, total domination in graphs. We have tended to primarily emphasize the recent selected results on total domination in graphs that appeared subsequent to the two domination books by Haynes, Hedetniemi, and Slater.

In this book, we do assume that the reader is acquainted with the basic concepts of graph theory and has had some exposure to graph theory at an introductory level. However, since graph theory terminology sometimes varies, the book is self-contained, and we clarify the terminology that will be adopted in this book in the introductory chapter.

We have written this book primarily to reach the following audience. The first audience is the graduate students who are interested in exploring the field of total domination theory in graphs and wish to "familiarize themselves with the subject, the research techniques, and the major accomplishments in the field." It is our hope that such graduate students will find topics and problems that can be developed further. The second audience is the established researcher in domination theory who wishes to have easy access to known results and latest developments in the field of total domination theory in graphs.

We have supplied several proofs for the reader to acquaint themselves with a toolbox of proof techniques and ideas with which to attack open problems in the field. We have identified many unsolved problems in the area and provided a chapter devoted to conjectures and open problems.

Chapter 1 introduces graph theory and hypergraph theory concepts fundamental to the chapters that follow. Perhaps much of the recent interest in total domination in graphs arises from the fact that total domination in graphs can be translated to the problem of finding transversals in hypergraphs. In this introductory chapter, we discuss the transition from total domination in graphs to transversals in hypergraphs.

Chapter 2 establishes fundamental properties of total dominating sets in graphs. General bounds relating the total domination number to other parameters are presented in this chapter, and properties of minimal total dominating sets are listed.

Complexity and algorithmic results on total domination in graphs are discussed in Chap. 3. We outline a few of the best-known algorithms and state what is currently known in this field. A linear algorithm to compute the total domination of a tree is given. We discuss fixed parameter tractability problems for the total domination number. A simple heuristic that finds a total dominating set in a graph is presented.

In Chap. 4 we present results on total domination in trees, the simplest class of graphs. A constructive characterization of trees with largest possible total domination in terms of the order and number of leaves of the tree is provided. We explore the relationship between the total domination number and the ordinary domination number of a tree.

In Chap. 5, we determine upper bounds on the total domination number of a graph in terms of its minimum degree. A general bound involving the minimum degree is presented. We then focus in detail on upper bounds on the total domination number when the minimum degree is one, two, three, four, or five, respectively, and consider each case in turn. Known upper bounds on the total domination number of a graph in terms of its order and minimum degree are summarized in an appropriate table. We close this chapter with a discussion of bounds on the total domination number of a graph with certain connectivity requirements imposed on the graph.

In Chap. 6, we turn our attention to investigate upper bounds on the total domination number of a planar graph of small diameter.

Chapter 7 determines whether the absence of any specified cycle guarantees that the upper bound on the total domination number established in Chap. 5 can be lowered. Upper bounds on the total domination number of a graph with given girth are presented.

In Chap. 8 we relate the size and the total domination number of a graph of given order. A linear Vizing-like relation is established relating the size of a graph and its order, total domination number, and maximum degree.

In Chap. 9 we impose the structural restriction of claw-freeness on a graph and investigate upper bounds on the total domination number of such graphs, while in Chap. 10, we discuss a relationship between the total domination number and the matching number of a graph.

An in-depth study of criticality issues relating to the total domination is presented in Chap. 11. In this chapter, we discuss an important association with total domination edge critical graphs and diameter critical graphs.

In Chap. 12 we investigate the behavior of the total domination number on a graph product. In particular, we focus our attention on the Cartesian product and the direct product of graphs.

Chapter 13 considers the problem of partitioning the vertex set of a graph into a total dominating set and something else. In this chapter, we exhibit a surprising connection between disjoint total dominating sets in graphs, 2 coloring of hypergraphs, and even cycles in digraphs.

In Chap. 14 we determine an upper bound of $1+\sqrt{n\ln(n)}$ on the total domination number of a graph with diameter two and show that this bound is close to optimum.

In Chap. 15 we present Nordhaus–Gaddum-type bounds for the total domination number of a graph. In Chap. 16 we focus our attention on the upper total domination number of a graph. By imposing a regularity condition on a graph, we show using edge weighting functions on total dominating sets how to obtain a sharp upper bound on the upper total domination of the graph.

In Chap. 17 we present various generalizations of the total domination number. We select four such variations of a total dominating set in a graph and briefly discuss each variation.

We close with Chap. 18 which lists several conjectures and open problems which have yet to be solved.

We wish to express our sincere thanks to Gary Chartrand, Teresa Haynes, and Justin Southey for their helpful comments, suggestions, and remarks which led to improvements in the presentation and enhanced the exposition of the manuscript. A special acknowledgement is due to Justin Southey for assisting us with some of the diagrams and for producing such superb figures.

We have tried to eliminate errors, but surely several remain. We do welcome any comments the reader may have. A list of typographical errors, corrections, and suggestions can be e-mailed to the authors.

Auckland Park, South Africa

Michael A. Henning
Anders Yeo

Contents

Chapter 1
Introduction

1.1 Introduction

Total domination in graphs was introduced by Cockayne, Dawes, and Hedetniemi [39] and is now well studied in graph theory. The literature on this subject has been surveyed and detailed in the two excellent so-called domination books by Haynes, Hedetniemi, and Slater [85, 86]. This book focuses primarily on recent selected results on total domination in graphs that appeared subsequent to the two domination books by Haynes, Hedetniemi, and Slater.

1.2 A Total Dominating Set in a Graph

We begin by defining a total dominating set in a graph. For this purpose, let G be a graph with vertex set $V(G)$ and edge set $E(G)$. The *open neighborhood* of a vertex $v \in V(G)$ is $N_G(v) = \{u \in V(G) \mid uv \in E(G)\}$ and its *closed neighborhood* is the set $N_G[v] = \{v\} \cup N_G(v)$. For a set $S \subseteq V$, its *open neighborhood* is the set $N(S) = \bigcup_{v \in S} N(v)$ and its *closed neighborhood* is the set $N[S] = N(S) \cup S$. If the graph G is clear from the context, we simply write V, E, $N(v)$ $N[v]$, $N(S)$, and $N[S]$ rather than $V(G)$, $E(G)$, $N_G(v)$, $N_G[v]$, $N_G(S)$, and $N_G[S]$, respectively.

A *total dominating set*, abbreviated TD-set, of a graph $G = (V, E)$ with no isolated vertex is a set S of vertices of G such that every vertex is adjacent to a vertex in S. Thus a set $S \subseteq V$ is a TD-set in G if $N(S) = V$. If no proper subset of S is a TD-set of G, then S is a *minimal TD-set* of G. Every graph without isolated vertices has a TD-set, since $S = V$ is such a set. The *total domination number* of G, denoted by $\gamma_t(G)$, is the minimum cardinality of a TD-set of G. A TD-set of G of cardinality $\gamma_t(G)$ is called a $\gamma_t(G)$-set. If X and Y are subsets of vertices in G, then the set X *totally dominates* the set Y in G, written $X \succ_t Y$, if $Y \subseteq N(X)$. In particular, if X totally dominates V, then X is a TD-set in G.

M.A. Henning and A. Yeo, *Total Domination in Graphs*, Springer Monographs in Mathematics, DOI 10.1007/978-1-4614-6525-6_1,

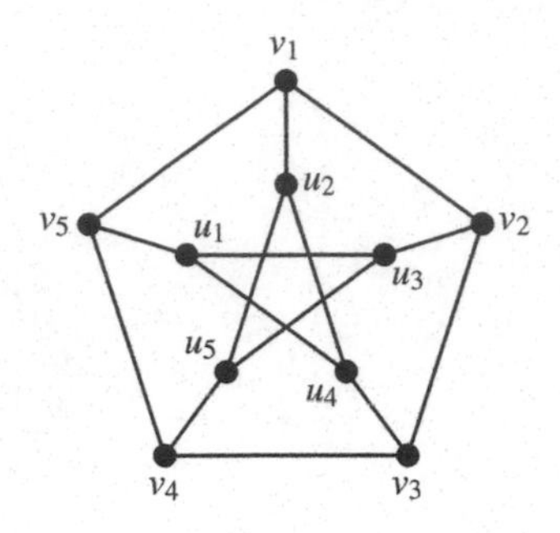

Fig. 1.1 The Petersen graphG_{10}

The *upper total domination number* of G, denoted by $\Gamma_t(G)$, is the maximum cardinality of a minimal TD-set in G. We call a minimal TD-set of cardinality $\Gamma_t(G)$ a $\Gamma_t(G)$-set.

To illustrate these definitions, consider the Petersen graph G_{10} as drawn in Fig. 1.1. The set $S = N[v_1] = \{v_1, u_2, v_2, v_5\}$ is a TD-set in G_{10}, and so $\gamma_t(G_{10}) \leq |S| = 4$. Since G_{10} is a cubic graph, each vertex totally dominates three vertices. Therefore no set of three or fewer vertices totally dominates all ten vertices in the graph, and so $\gamma(G_{10}) \geq 4$. Consequently, $\gamma_t(G_{10}) = 4$. One can in fact show that the Petersen graph has exactly ten distinct $\gamma_t(G_{10})$-sets, namely, the ten closed neighborhoods $N[v]$ corresponding to the ten vertices v in the graph. The sets $\{v_1, v_2, v_3, v_4, v_5\}$ and $\{v_1, v_3, v_5, u_3, u_4, u_5\}$ are both minimal TD-sets in the Petersen graph G_{10}. Hence, G_{10} contains minimal TD-sets of cardinalities 4, 5, and 6. One can further show that the Petersen graph has no minimal TD-set of cardinality exceeding 6, implying that $\Gamma_t(G_{10}) = 6$.

A *dominating set* of G is a set D of vertices of G such that every vertex in $V \setminus D$ is adjacent to a vertex in D. Hence, a set D is a dominating set of G if $N[v] \cap D \neq \emptyset$ for every vertex v in G, or, equivalently, $N[D] = V$. The *domination number* of G, denoted by $\gamma(G)$, is the minimum cardinality of a dominating set. A dominating set of G of cardinality $\gamma(G)$ is called a $\gamma(G)$-set. If X and Y are subsets of vertices in G, then the set X *dominates* the set Y in G, written $X \succ Y$, if $Y \subseteq N[X]$. In particular, if X dominates V, then X is a dominating set in G.

A TD-set in a graph is in some ways a more natural concept than a dominating set in the sense that total domination in graphs can be translated to the problem of finding transversals in hypergraphs as follows. For a graph G, let the open neighborhood hypergraph of G, denoted by $ONH(G)$, have vertex set $V(G)$ and edge set $\{N_G(x) \mid x \in V(G)\}$. A transversal in $ONH(G)$ is a set of vertices intersecting every edge of $ONH(G)$, which can be shown to be equivalent to a TD-set in G. We will describe this transition in detail in Sect. 1.6.

Using this interplay between total dominating sets in graphs and transversals in hypergraphs, several results on total domination in graphs can be obtained that appear very difficult to obtain using purely graph theoretic techniques. For example, it is relatively easy to prove a sharp upper bound on the total domination number of a connected cubic graph in terms of its order, while it remains one of the major outstanding problems to determine such a bound on the domination number. The main strength of this approach of using transversals in hypergraphs is that in the induction steps we can move to hypergraphs which are not open neighborhood hypergraphs of any graph, thereby giving us much more flexibility.

We remark that while domination in graphs can also be translated to the problem of finding transversals in hypergraphs, the results so obtained do not seem as effective and powerful as using a purely graph theoretic approach when attacking problems on the ordinary domination number.

In Sect. 1.3 we define graph theory terminology and concepts that we will need in subsequent chapters. The reader who is familiar with graph theory will no doubt be acquainted with the terminology in Sect. 1.3. However since graph theory terminology sometimes varies, we clarify the terminology that will be adopted in this book.

1.3 Graph Theory Terminology and Concepts

For notation and graph theory terminology, we in general follow [85]. Specifically, a *graph* G is a finite nonempty set $V(G)$ of objects called *vertices* (the singular is *vertex*) together with a possibly empty set $E(G)$ of 2-element subsets of $V(G)$ called *edges*. The *order* of G is $n(G) = |V(G)|$ and the size of G is $m(G) = |E(G)|$. The *degree* of a vertex $v \in V(G)$ in G is $d_G(v) = |N_G(v)|$. A vertex of degree one is called an *end-vertex*. The minimum and maximum degree among the vertices of G is denoted by $\delta(G)$ and $\Delta(G)$, respectively. Further for a subset $S \subseteq V(G)$, the *degree of v in S*, denoted $d_S(v)$, is the number of vertices in S adjacent to v; that is, $d_S(v) = |N(v) \cap S|$. In particular, $d_G(v) = d_V(v)$. If the graph G is clear from the context, we simply write V, E, n, m, $d(v)$, δ, and Δ rather than $V(G)$, $E(G)$, $n(G)$, $m(G)$, $d_G(v)$, $\delta(G)$, and $\Delta(G)$, respectively. To indicate that a graph G has vertex set V and edge set E we write $G = (V,E)$.

If G is a graph, the complement of G, denoted by $\overline{G}$, is formed by taking the vertex set of G and joining two vertices by an edge whenever they are not joined in G. If H is a supergraph of G, then the graph $H - E(G)$ is called the *complement of G relative to H*. Further, if $H = K_{s,s}$ for some $s \geq 2$ and G is a spanning subgraph of H, then we also call the complement of G relative to H the *bipartite complement of G*.

Let $S \subseteq V$ and let v be a vertex in S. The *S-private neighborhood* of v is defined by $\text{pn}[v,S] = \{w \in V \mid N_G[w] \cap S = \{v\}\}$, while its *open S-private neighborhood* is defined by $\text{pn}(v,S) = \{w \in V \mid N_G(w) \cap S = \{v\}\}$. We remark that the sets $\text{pn}[v,S] \setminus S$ and $\text{pn}(v,S) \setminus S$ are equivalent, and we define the *S-external private neighborhood* of v to be this set, abbreviated $\text{epn}[v,S]$ or $\text{epn}(v,S)$. The *S-internal private neighborhood* of v is defined by $\text{ipn}[v,S] = \text{pn}[v,S] \cap S$ and its *open S-internal private neighborhood* is defined by $\text{ipn}(v,S) = \text{pn}(v,S) \cap S$. We define an *$S$-external private neighbor* of v to be a vertex in $\text{epn}(v,S)$ and an *S-internal private neighbor* of v to be a vertex in $\text{ipn}(v,S)$.

For subsets $X,Y \subseteq V$, we denote the set of edges that join a vertex of X and a vertex of Y by $[X,Y]$. Thus, $|[X,Y]|$ is the number of edges with one end in X and the other end in Y. In particular, $|[X,X]| = m(G[X])$.

For a set $S \subseteq V$, the subgraph induced by S is denoted by $G[S]$. A *cycle* and *path* on n vertices are denoted by C_n and P_n, respectively. We denote the *complete bipartite graph* with partite sets of cardinality n and m by $K_{n,m}$. We denote the graph obtained from $K_{n,m}$ by deleting one edge by $K_{n,m} - e$. The *girth* $g(G)$ of G is the length of a shortest cycle in G. If G is a disjoint union of k copies of a graph F, we write $G = kF$.

For two vertices u and v in a connected graph G, the *distance* $d(u,v)$ between u and v is the length of a shortest (u,v)-path in G. The maximum distance among all pairs of vertices of G is the *diameter* of G, which is denoted by $\text{diam}(G)$. We say that G is a *diameter-2 graph* if $\text{diam}(G) = 2$.

The *eccentricity* of a vertex v in a connected graph G is the maximum of the distances from v to the other vertices of G and is denoted by $\text{ecc}_G(v)$. Another definition of the diameter is the maximum eccentricity taken over all vertices of G. The minimum eccentricity taken over all vertices of G is called the *radius* of G and is denoted by $\text{rad}(G)$. The distance $d_G(v,S)$ between a vertex v and a set S of vertices in a graph G is the minimum distance from v to a vertex of S in G. The *eccentricity* denoted by $\text{ecc}_G(S)$ of S in G is the maximum distance between S and a vertex farthest from S in G; that is, $\text{ecc}_G(S) = \max\{d_G(v,S) \mid v \in V(G)\}$. In particular, we note that if $v \in S$, then $d_G(v,S) = 0$, implying that if $S = V(G)$, then $\text{ecc}_G(S) = 0$.

If $G_1, G_2, \ldots, G_k$ are graphs on the same vertex set but with pairwise disjoint edge sets, then $G_1 \oplus G_2 \oplus \cdots \oplus G_k$ denotes the graph whose edge set is the union of their edge sets. In particular, if G and H are graphs on the same vertex set but with disjoint edge sets, then $G \oplus H$ denotes the graph whose edge set is the union of their edge sets.

If G does not contain a graph F as an induced subgraph, then we say that G is *F-free*. In particular, we say a graph is *claw-free* if it is $K_{1,3}$-free and *diamond-free* if it is $(K_4 - e)$-free, where $K_4 - e$ denotes the complete graph on four vertices minus one edge.

The graph G is k-regular if $d(v) = k$ for every vertex $v \in V$. A *regular graph* is a graph that is k-regular for some $k \geq 0$. A 3-regular graph is also called a cubic graph.

For graphs G and H, the *Cartesian product* $G \square H$ is the graph with vertex set $V(G) \times V(H)$ where two vertices (u_1, v_1) and (u_2, v_2) are adjacent if and only if either $u_1 = u_2$ and $v_1 v_2 \in E(H)$ or $v_1 = v_2$ and $u_1 u_2 \in E(G)$.

A subset S of vertices in a graph G is a *packing* (respectively, an *open packing*) if the closed (respectively, open) neighborhoods of vertices in S are pairwise disjoint. The *packing number* $\rho(G)$ is the maximum cardinality of a packing, while the *open packing number* $\rho^o(G)$ is the maximum cardinality of an open packing. A set S of vertices in G is an *independent set* (also called a *stable set* in the literature) if no two vertices of S are adjacent in G. The size of a maximum independent set in G is denoted by $\alpha(G)$.

Two edges in a graph G are *independent* if they are not adjacent in G. A *matching* in a graph G is a set of independent edges in G, while a matching of maximum cardinality is a *maximum matching*. The number of edges in a maximum matching of G is called the *matching number* of G which we denote by $\alpha'(G)$. A *perfect matching* M in G is a matching in G such that every vertex of G is incident to an edge of M. Thus, G has a perfect matching if and only if $\alpha'(G) = |V(G)|/2$. The graph G is *factor-critical* if $G - v$ has a perfect matching for every vertex v in G.

An *odd component* of a graph G is a component of odd order. The number of odd components of G is denoted by $\mathrm{oc}(G)$.

A *rooted tree* distinguishes one vertex r called the *root*. For each vertex $v \neq r$ of T, the *parent* of v is the neighbor of v on the unique (r, v)-path, while a *child* of v is any other neighbor of v. A *descendant* of v is a vertex u such that the unique (r, u)-path contains v. Thus, every child of v is a descendant of v. We let $C(v)$ and $D(v)$ denote the set of children and descendants, respectively, of v, and we define $D[v] = D(v) \cup \{v\}$. The *maximal subtree* at v is the subtree of T induced by $D[v]$ and is denoted by T_v. A leaf of T is a vertex of degree 1, while a support vertex of T is a vertex adjacent to a leaf. A *branch vertex* is a vertex of degree at least 3 in T.

Let H be a graph. The *corona* $H \circ K_1$ of H, also denoted $\mathrm{cor}(H)$ in the literature, is the graph obtained from H by adding a pendant edge to each vertex of H. The 2-*corona* $H \circ P_2$ of H is the graph of order $3|V(H)|$ obtained from H by attaching a path of length 2 to each vertex of H so that the resulting paths are vertex disjoint.

1.4 Digraph Terminology and Concepts

Let $D = (V, A)$ be a digraph with vertex set V and arc set A, and let v be a vertex of D. The *outdegree* $d^+(v)$ of v is the number of arcs leaving v; that is, $d^+(v)$ is the number of arcs of the form (v, u). The *indegree* $d^-(v)$ of v is the number of arcs entering v. Let $\delta^+(D)$ and $\delta^-(D)$ denote the minimum outdegree and indegree, respectively, in D. The *closed out-neighborhood* of v is the set $N^+[v]$ consisting of v and all its out-neighbors. The digraph D is *k-regular* if $d^+(v) = d^-(v) = k$ for every vertex $v \in V$. By a *path* (respectively, *cycle*) in D, we mean a directed path (respectively, directed cycle).

1.5 Hypergraph Terminology and Concepts

Hypergraphs are systems of sets which are conceived as natural extensions of graphs. A *hypergraph* $H = (V, E)$ is a finite nonempty set $V = V(H)$ of elements, called *vertices*, together with a finite multiset $E = E(H)$ of subsets of V, called *hyperedges* or simply *edges*. The *order* and *size* of H are $n = |V|$ and $m = |E|$, respectively. A *k-edge* in H is an edge of size k. The hypergraph H is said to be *k-uniform* if every edge of H is a k-edge. The *rank* of H is the size of a largest edge in H. Every (simple) graph is a 2-uniform hypergraph. Thus graphs are special hypergraphs.

The *degree* of a vertex v in H, denoted by $d_H(v)$ or simply by $d(v)$ if H is clear from the context, is the number of edges of H which contain v. The minimum and maximum degrees among the vertices of H are denoted by $\delta(H)$ and $\Delta(H)$, respectively.

Two vertices x and y of H are *adjacent* if there is an edge e of H such that $\{x,y\} \subseteq e$. The *neighborhood* of a vertex v in H, denoted $N_H(v)$ or simply $N(v)$ if H is clear from the context, is the set of all vertices different from v that are adjacent to v. Two vertices x and y of H are *connected* if there is a sequence $x = v_0, v_1, v_2 \dots, v_k = y$ of vertices of H in which v_{i-1} is adjacent to v_i for $i = 1, 2, \dots, k$. A *connected hypergraph* is a hypergraph in which every pair of vertices are connected. A maximal connected subhypergraph of H is a *component* of H. Thus, no edge in H contains vertices from different components. A hypergraph H is called an *intersecting hypergraph* if every two distinct edges of H have a nonempty intersection.

A subset T of vertices in a hypergraph H is a *transversal* (also called *vertex cover* or *hitting set* in many papers) if T has a nonempty intersection with every edge of H. The *transversal number* $\tau(H)$ of H is the minimum size of a transversal in H. A $\tau(H)$-transversal is a transversal in H of size $\tau(H)$. We say that an edge e in H is *covered* by a set T if $e \cap T \neq \emptyset$. In particular, if T is a transversal in H, then T covers every edge of H.

For a subset $X \subset V(H)$ of vertices in H, we define $H - X$ to be the hypergraph obtained from H by deleting the vertices in X and all edges incident with X. Note that if T' is a transversal in $H - X$, then $T' \cup X$ is a transversal in H.

1.6 The Transition from Total Domination in Graphs to Transversals in Hypergraphs

In this section, we discuss the transition from total domination in graphs to transversals in hypergraphs. For a graph $G = (V,E)$, recall that $ONH(G)$, or simply H_G for notational convenience, is the open neighborhood hypergraph (abbreviated *ONH*) of G; that is, $H_G = (V,C)$ is the hypergraph with vertex set $V(H_G) = V$ and with edge set $E(H_G) = C = \{N_G(x) \mid x \in V\}$ consisting of the open neighborhoods of vertices of V in G. For example, consider the generalized Petersen graph G_{16} shown in Fig. 1.2.

The *ONH* of the generalized Petersen graph G_{16} is shown in Fig. 1.3.

In general, one can show the following result, a proof of which is given in [133].

Theorem 1.1 ([133]). *The ONH of a connected bipartite graph consists of two components, while the ONH of a connected graph that is not bipartite is connected.*

The following result shows that the transversal number of the *ONH* of a graph is precisely the total domination number of the graph.

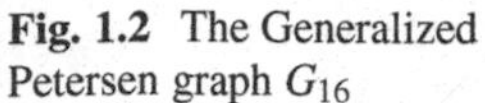

Fig. 1.2 The Generalized Petersen graph G_{16}

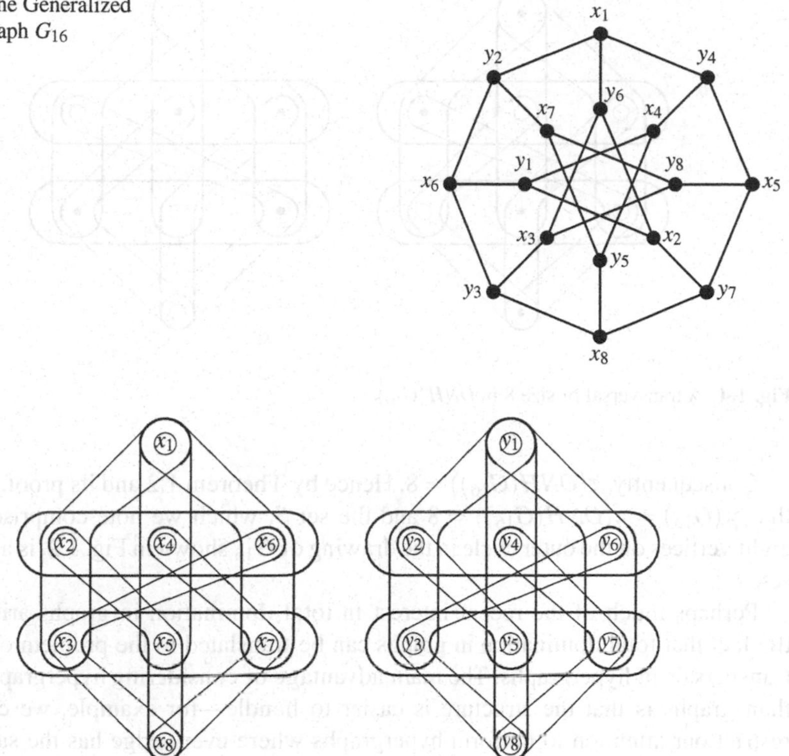

Fig. 1.3 The *ONH* of G_{16}

Theorem 1.2. *If G is a graph with no isolated vertex and H_G is the ONH of G, then $\gamma_t(G) = \tau(H_G)$.*

Proof. On the one hand, every TD-set in G contains a vertex from the open neighborhood of each vertex in G and is therefore a transversal in H_G. On the other hand, every transversal in H_G contains a vertex from the open neighborhood of each vertex of G and is therefore a TD-set in G. So $\gamma_t(G) = \tau(H_G)$. □

To illustrate Theorem 1.2, let us consider the generalized Petersen graph G_{16} shown in Fig. 1.2. The *ONH* of G_{16} consists of the two isomorphic components shown in Fig. 1.3. Each component requires four vertices to cover all its hyperedges, and so $\tau(ONH(G_{16})) \geq 8$. However the set $S = \{x_1, x_5, x_6, x_8, y_2, y_3, y_4, y_7\}$ is an example of a transversal of cardinality 8 in $ONH(G_{16})$, as indicated by the eight darkened vertices in Fig. 1.4 (where the labels of the vertices have been omitted), and so $\tau(ONH(G_{16})) \leq 8$.

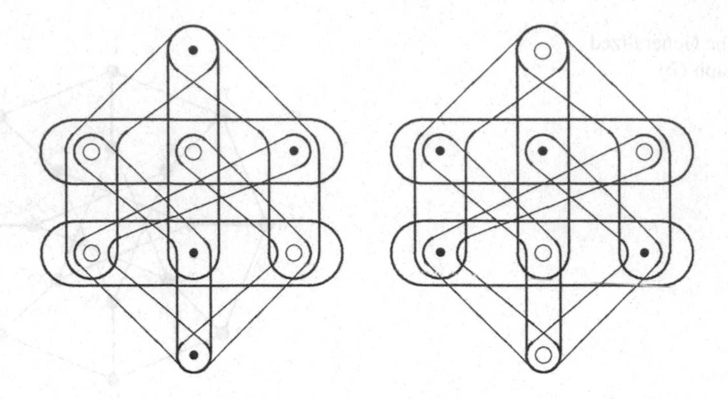

Fig. 1.4 A transversal of size 8 in $ONH(G_{16})$

Consequently, $\tau(ONH(G_{16})) = 8$. Hence by Theorem 1.2 and its proof, we have that $\gamma_t(G_{16}) = \tau(ONH(G_{16})) = 8$ and the set S, which we note comprises of the eight vertices on the outer cycle in the drawing of G_{16} shown in Fig. 1.2, is a $\gamma_t(G_{16})$-set.

Perhaps much of the recent interest in total domination in graphs arises from the fact that total domination in graphs can be translated to the problem of finding transversals in hypergraphs. The main advantage of considering hypergraphs rather than graphs is that the structure is easier to handle—for example, we can often restrict our attention to uniform hypergraphs where every edge has the same size. Furthermore when using induction we can move to hypergraphs which are not *ONH*s of any graph, giving much more flexibility than just using graphs. Also, we do not lose much information by going to hypergraphs, as any hypergraph may appear as a component of some *ONH*, which is not the case when we consider domination (where we would need to consider the closed neighborhood hypergraph) instead of total domination. This is the reason that considering transversals in hypergraphs seems to be a much more powerful tool for total domination than for domination. The idea of using transversals in hypergraphs to obtain results on total domination in graphs first appeared in a paper by Thomassé and Yeo [197] submitted in 2003. Up to that time, the transition from total domination in graphs to transversals in hypergraphs seemed to pass by unnoticed. Subsequent to the Thomassé–Yeo paper, several important results on total domination in graphs have been obtained using transversals in hypergraphs.

Chapter 2
Properties of Total Dominating Sets and General Bounds

2.1 Introduction

In order to obtain results on the total domination number, we need to first establish properties of TD-sets in graphs. In this chapter, we list properties of minimal TD-sets in a graph. Further we present general bounds relating the total domination number to other parameters.

2.2 Properties of Total Dominating Sets

Recall that if $G = (V,E)$ is a graph, $S \subseteq V$, and $v \in S$, then $\text{pn}(v,S) = \{w \in V \mid N(w) \cap S = \{v\}\}$, $\text{ipn}(v,S) = \text{pn}(v,S) \cap S$ and $\text{epn}(v,S) = \text{pn}(v,S) \setminus S$. The following property of a minimal TD-set in a graph is established by Cockayne, Dawes, and Hedetniemi [39].

Proposition 2.1 ([39]). *Let S be a TD-set in a graph G. Then, S is a minimal TD-set in G if and only if $|\text{epn}(v,S)| \geq 1$ or $|\text{ipn}(v,S)| \geq 1$ for each $v \in S$.*

Proof. Let S be a minimal TD-set in G and let $v \in S$. If $|\text{epn}(v,S)| = 0$ and $|\text{ipn}(v,S)| = 0$, then every vertex $x \in V(G)$ must be adjacent to a vertex in $S \setminus \{v\}$ as $N(x) \cap S \neq \{v\}$. Hence, $S \setminus \{v\}$ is a TD-set of G, contradicting the minimality of S. Therefore, $|\text{epn}(v,S)| \geq 1$ or $|\text{ipn}(v,S)| \geq 1$ for each $v \in S$. Conversely, if $|\text{epn}(v,S)| \geq 1$ or $|\text{ipn}(v,S)| \geq 1$ for each $v \in S$, then $S \setminus \{v\}$ is not a TD-set, implying that S is a minimal TD-set in G. □

The following stronger property of a minimum TD-set in a graph is established in [102].

Theorem 2.2 ([102]). *If G is a connected graph of order $n \geq 3$ and $G \not\cong K_n$, then G has a minimum TD-set S such that every vertex $v \in S$ satisfies $|\text{epn}(v,S)| \geq 1$ or is adjacent to a vertex v' of degree 1 in $G[S]$ satisfying $|\text{epn}(v',S)| \geq 1$.*

M.A. Henning and A. Yeo, *Total Domination in Graphs*, Springer Monographs in Mathematics, DOI 10.1007/978-1-4614-6525-6_2,

For a subset S of vertices in a graph G, the *open boundary* of S is defined as $OB(S) = \{v\colon |N(v) \cap S| = 1\}$; that is, $OB(S)$ is the set of vertices totally dominated by exactly one vertex in S. Hedetniemi, Jacobs, Laskar, and Pillone characterized a minimal TD-set by its open boundary as follows (see Theorem 6.10 in [85]).

Theorem 2.3 ([85]). *A TD-set S in a graph G is a minimal TD-set if and only if $OB(S)$ totally dominates S.*

Proof. Suppose first that $OB(S)$ totally dominates S. Let $v \in S$ and let u be a vertex in $OB(S)$ that is adjacent to v. Then, $N(u) \cap S = \{v\}$. If $u \notin S$, then $u \in \text{epn}(v,S)$. If $u \in S$, then $u \in \text{ipn}(v,S)$. Hence, $\text{epn}(v,S) \neq \emptyset$ or $\text{ipn}(v,S) \neq \emptyset$ for every vertex $v \in S$. Thus, by Proposition 2.1, S is a minimal TD-set. To prove the necessity, suppose that S is a minimal TD-set. Let $v \in S$. By Proposition 2.1, $\text{epn}(v,S) \neq \emptyset$ or $\text{ipn}(v,S) \neq \emptyset$. If $\text{epn}(v,S) \neq \emptyset$, then there exists a vertex $u \in V \setminus S$ such that $N(u) \cap S = \{v\}$, and so $u \in OB(S)$. On the other hand, if $\text{ipn}(v,S) \neq \emptyset$, then there exists a vertex $u \in S$ such that $N(u) \cap S = \{v\}$, and so $u \in OB(S)$. In both cases, $OB(S)$ totally dominates the vertex v. □

A graph class is *hereditary* if it is closed under taking induced subgraphs and *additive* if it is closed under a disjoint union of graphs. An additive hereditary graph class is said to be *nontrivial* if it is nonempty and contains K_2. We note that triangle-free graphs or bipartite graphs are examples of a nontrivial additive hereditary graph class. Further we note that the minimal forbidden subgraphs of an additive hereditary graph class are connected. For example the 3-cycle is the only minimal forbidden subgraph of triangle-free graphs, while all odd cycles are the minimal forbidden subgraphs of bipartite graphs.

Using results of Bacsó [9] and Tuza [201] who independently gave a full characterization of the graphs for which every connected induced subgraph has a connected dominating subgraph satisfying an arbitrary prescribed hereditary property, Schaudt [180] derived a similar characterization of the graphs for which any isolate-free induced subgraph has a TD-set that satisfies a prescribed additive hereditary property. If $\mathscr{G}$ is a graph class, then following the notation of Schaudt, we denote by $Total(\mathscr{G})$ the set of isolate-free graphs for which every isolate-free subgraph H has a TD-set that is isomorphic to some member of $\mathscr{G}$. Restricting his attention to graph classes that are hereditary and additive, Schaudt [180] characterized $Total(\mathscr{G})$ in terms of minimal forbidden subgraphs, for arbitrary nontrivial additive hereditary properties $\mathscr{G}$. The following result shows that the corona graphs are the only minimal forbidden subgraphs of $Total(\mathscr{G})$, where recall that a corona graph $H \circ K_1$ is a graph that can be obtained from a graph H by adding a pendant edge to each vertex of H.

Theorem 2.4 ([180]). *Let $\mathscr{G}$ be a nontrivial additive hereditary graph class containing all paths. Then the minimal forbidden subgraphs of $Total(\mathscr{G})$ are the corona graphs of the minimal forbidden subgraphs of $\mathscr{G}$.*

A characterization of additive hereditary graph classes $\mathscr{G}$ which do not contain all paths remains an open problem. The following partial characterizations and sufficient conditions are given by Schaudt [180]. For any $k \geq 3$ and $2 \leq i \leq k-1$, let T_k^i be the graph obtained from the path P_k by attaching a pendant vertex to the i-th vertex of P_k and let $\mathscr{T}_k = \{T_k^i \mid 2 \leq i \leq k-1\}$ be the collection of these graphs.

Theorem 2.5 ([180]). *Let $\mathscr{G}$ be a nontrivial additive hereditary graph class that does not contain all paths and let k be minimal such that $P_k \notin \mathscr{G}$. Then the following hold:*

(a) *If $k = 3$, then the minimal forbidden subgraphs of $Total(\mathscr{G})$ are C_5 and the coronas of the minimal forbidden subgraphs of $\mathscr{G}$.*
(b) *If $k \geq 4$ and $\mathscr{G} \cap \mathscr{T}_{k-1} \neq \emptyset$, then the minimal forbidden subgraphs of $Total(\mathscr{G})$ are the coronas of the minimal forbidden subgraphs of $\mathscr{G}$.*
(c) *If $k \geq 4$, then $Total(\mathscr{G})$ contains all graphs that do not contain a corona of the minimal forbidden subgraphs of $\mathscr{G}$ as subgraph and do not contain any graph of $\{C_i \mid 5 \leq i \leq k+2\} \cup \{P_{k-1} \circ K_1\}$ as subgraph.*

2.3 General Bounds

In this section we present general bounds relating the total domination number to other parameters.

2.3.1 Bounds in Terms of the Order

We begin with the following bound on the total domination number of a connected graph in terms of the order of the graph due to Cockayne et al. [39].

Theorem 2.6 ([39]). *If G is a connected graph of order $n \geq 3$, then $\gamma_t(G) \leq 2n/3$.*

Brigham, Carrington, and Vitray [20] characterized the connected graphs of order at least 3 with total domination number exactly two-thirds their order.

Theorem 2.7 ([20]). *Let G be a connected graph of order $n \geq 3$. Then $\gamma_t(G) = 2n/3$ if and only if G is C_3, C_6, or $F \circ P_2$ for some connected graph F.*

Since Theorems 2.6 and 2.7 are fundamental results on total domination, we present a proof of these two results. The first proof we present is a graph theory proof from [130] and uses the property of minimal TD-sets in Sect. 2.2. We will afterwards give a hypergraph proof of Theorem 2.6.

Proof. Let $G = (V,E)$ be a connected graph of order $n \geq 3$. If $G = K_n$, then $\gamma_t(G) = 2 \leq 2n/3$. Further, if $\gamma_t(G) = 2n/3$, then $G = K_3$. Hence we may assume that $G \neq K_n$. Let S be a $\gamma_t(G)$-set satisfying the statement of Theorem 2.2.

Let $A = \{v \in S \mid \text{epn}(v,S) = \emptyset\}$ and let $B = S \setminus A$. By Theorem 2.2, each vertex $v \in A$ is adjacent to at least one vertex of B which is adjacent to v but to no other vertex of S. Hence, $|S| = |A| + |B| \leq 2|B|$. Let C be the set of all external S-private neighbors. Then, $C \subseteq V \setminus S$. Since each vertex of B has at least one external S-private neighbor, $|C| \geq |B|$. Hence,

$$n - |S| = |V \setminus S| \geq |C| \geq |B| \geq |S|/2, \tag{2.1}$$

and so $\gamma_t(G) = |S| \leq 2n/3$. This proves Theorem 2.6.

To prove Theorem 2.7, suppose that $\gamma_t(G) = 2n/3$ (and still $G \neq K_n$). Then we must have equality throughout the inequality chain (2.1). In particular, $|A| = |B| = |C|$ and $V \setminus S = C$. We deduce that each vertex of B therefore has degree 2 in G and is adjacent to a unique vertex of A and a unique vertex of C. Hence, G contains a spanning subgraph that consists of r disjoint copies of a path P_3 on three vertices, where $r = n/3$

On the one hand, if both sets A and C contain a vertex of degree 2 or more in G, then the connected graph G contains a spanning subgraph H where $H = C_6$ or $H = P_9 \cup (r-3)P_3$. If $H = P_9 \cup (r-3)P_3$, then $\gamma_t(H) = 5 + 2(r-3) = 2r - 1 < 2n/3$. Since adding edges to a graph does not increase its total domination number, $\gamma_t(G) \leq \gamma_t(H) < 2n/3$, contrary to our assumption. Hence, $H = C_6$. But then G can contain no additional edges not in H, and so $G = H = C_6$. Hence in this case, $G = C_6$.

On the other hand, if every vertex of A has degree 1 in G, then by the connectivity of G, the subgraph $G[C]$ is connected and $G = F \circ P_2$ where $F = G[C]$, while if every vertex of C has degree 1 in G, then the subgraph $G[A]$ is connected and $G = F \circ P_2$ where $F = G[A]$. This establishes Theorem 2.7. □

The second proof of Theorem 2.6 that we present is the analogous hypergraph proof of this result. This second proof serves to gently introduce the reader who may not be familiar with hypergraphs to the transition from total domination in graphs to transversals in hypergraphs. We shall first need the following hypergraph result which was proven independently by several authors (see, e.g., [35] for a more general result).

Theorem 2.8. *If $H = (V,E)$ is a connected hypergraph with all edges of size at least two, then $3\tau(H) \leq |V| + |E|$.*

Proof. We proceed by induction on the number $n = |V| \geq 1$ of vertices. Let H have size $m = |E|$. If $m = 0$, then $\tau(H) = 0$ and the theorem holds. Therefore the theorem holds when $n = 1$. Assume, then, that $n \geq 2$ and that the statement holds for all connected hypergraphs $H = (V', E')$ with all edges of size at least two where $|V'| < n$. Let $H = (V,E)$ be a connected hypergraph with all edges of size at least two where $|V| = n$ and $|E| = m$.

If $\Delta(H) = 1$, then $m = 1$, $n \geq 2$ and $\tau(H) = 1$, implying that the theorem holds. Hence, we may assume that $\Delta(H) \geq 2$. Let v be a vertex of maximum degree in H, and so $d_H(v) = \Delta(H)$. Let $H' = H - v$. Then, $H' = (V', E')$ is a hypergraph of order $|V'| = n - 1$ and size $|E'| \leq m - 2$ with all edges of size at least two. Let T' be a $\tau(H')$-transversal in H'. Applying the induction hypothesis to each component of

H', we have that $3\tau(H') = 3|T'| \le |V'| + |E'| \le |V| + |E| - 3$. Since $T = T' \cup \{v\}$ is a transversal of H, we have that $3\tau(H) \le 3|T| = 3(|T'| + 1) \le |V| + |E|$, establishing the desired upper bound. □

Recall that a 2-uniform hypergraph is a graph and that a complete graph is a graph in which every two vertices is adjacent. We remark that with a bit more work one can show that equality holds in Theorem 2.8 if and only if H is a complete graph on two or three vertices. We omit the details. We next present our second proof of Theorem 2.6.

Proof (Analogous hypergraph proof of Theorem 2.6). Let $G = (V,E)$ be a connected graph of order $n \ge 3$. We first consider the case when $\delta(G) \ge 2$. Let H_G be the *ONH* of G, and so $n(H_G) = m(H_G) = n(G) = n$. By Theorems 1.2 and 2.8, $\gamma_t(G) = \tau(H_G) \le (n(H_G) + m(H_G))/3 = 2n/3$, which establishes the desired upper bound. Therefore we may assume that $\delta(G) = 1$.

Let X be the set of all vertices of degree one in G, and so $X = \{v \in V(G) \mid d_G(v) = 1\}$. By assumption $\delta(G) = 1$, and so $|X| \ge 1$. Since G is a connected graph on at least three vertices, the set X is an independent set in G. Let $Y = N(X)$ and let $Z = N(Y) \setminus (X \cup Y)$. Then, $|X| \ge |Y|$. We now consider the following two cases:

Case 1. $|Z| < |Y|$*:* If $V(G) = X \cup Y \cup Z$, then $Y \cup Z$ is a TD-set of G. Further since $|X| \ge |Y|$ and $|Y| > |Z|$, we have that $2n = 2|X| + 2|Y| + 2|Z| > 3|Y| + 3|Z|$, and so $\gamma_t(G) \le |Y| + |Z| < 2n/3$. Hence we may assume that $V(G) \ne X \cup Y \cup Z$, for otherwise the desired result follows. We now consider the hypergraph $H = H_G - (X \cup Y \cup Z)$, and so H is obtained from H_G by removing all vertices in $X \cup Y \cup Z$ and all edges that intersect $X \cup Y \cup Z$. Applying Theorem 2.8 to every component of H, we get $\tau(H) \le (n(H) + m(H))/3$. Every $\tau(H)$-set can be extended to a transversal in H_G by adding to it the set $Y \cup Z$. Hence by Theorem 1.2 and the fact that $|Z| < |Y|$ and $|Y| \le |X|$, we get the following, which completes the proof of Case 1:

$$\begin{aligned}
\gamma_t(G) &= \tau(H_G) \\
&\le |Y| + |Z| + \tau(H) \\
&\le |Y| + |Z| + \left(\tfrac{n(H)+m(H)}{3}\right) \\
&\le |Y| + |Z| + \left(\tfrac{n(H_G)-|X|-|Y|-|Z|}{3}\right) + \left(\tfrac{m(H_G)-|X|-|Y|-|Z|}{3}\right) \\
&= \left(\tfrac{n(H_G)+m(H_G)}{3}\right) + \left(\tfrac{|Y|+|Z|-2|X|}{3}\right) \\
&< \tfrac{2n}{3}.
\end{aligned}$$

Case 2. $|Z| \ge |Y|$*:* Let $H = H_G - Y$, and so H is obtained from H_G by removing all vertices in Y and all edges that intersect Y. We note that there are at least $|X| + |Z|$ edges in H_G that intersect Y, and so $m(H) \le m(H_G) - |X| - |Z| \le m(H_G) - 2|Y| = n - 2|Y|$. Further, $n(H) = n(H_G) - |Y| = n - |Y|$, and so $n(H) + m(H) \le 2n - 3|Y|$. Every $\tau(H)$-set can be extended to a transversal in H_G by adding to it the set Y. Hence by Theorems 1.2 and 2.8, we have that $\gamma_t(G) = \tau(H_G) \le |Y| + \tau(H) \le |Y| + (n(H) + m(H))/3 \le |Y| + (2n - 3|Y|)/3 = 2n/3$, which establishes the desired result. □

We remark that with a bit more work the above hypergraph proof of Theorem 2.6 can be expanded to give a hypergraph proof of Theorem 2.7. We omit the details. A more detailed discussion of bounds on the total domination number of a graph in terms of its order is given in Chap. 5.

2.3.2 Bounds in Terms of the Order and Size

The total domination number of a cycle C_n or path P_n on $n \geq 3$ vertices is easy to compute.

Observation 2.9. *For* $n \geq 3$, $\gamma_t(P_n) = \gamma_t(C_n) = \lfloor n/2 \rfloor + \lceil n/4 \rceil - \lfloor n/4 \rfloor$. *In other words,*

$$\gamma_t(P_n) = \gamma_t(C_n) = \begin{cases} \frac{n}{2} & \text{if } n \equiv 0 \pmod 4 \\ \frac{n+1}{2} & \text{if } n \equiv 1,3 \pmod 4 \\ \frac{n}{2}+1 & \text{if } n \equiv 2 \pmod 4 \end{cases}$$

In this section, we therefore restrict our attention to graphs with maximum degree at least 3. The following upper bound on the total domination number of a graph in terms of both its order and size is given in [106,184].

Theorem 2.10 ([106,184]). *If G is a connected graph with* $\Delta(G) \leq 3$ *and of order n and size m, then* $\gamma_t(G) \leq n - m/3$.

Note that if G is a cubic graph of order n and size m, then $2m = \sum_{v \in V(G)} d(v) = 3n$, which by Theorem 2.10 implies that $\gamma_t(G) \leq n/2$. In Sect. 5.5 we will generalize this result on cubic graphs to minimum degree three graphs. A more detailed discussion of bounds on the total domination number of a graph in terms of both its order and size is given in Chap. 8.

2.3.3 Bounds in Terms of Maximum Degree

The following result provides a trivial lower bound on the total domination number of a graph in terms of the maximum degree of the graph.

Theorem 2.11. *If G is a graph of order n with no isolated vertex, then* $\gamma_t(G) \geq n/\Delta(G)$.

Proof. Let S be a $\gamma_t(G)$-set. Since every vertex is totally dominated by the set S, every vertex belongs to the open neighborhood of at least one vertex in S. Hence, $V(G) = \cup_{v \in S} N_G(v)$, implying that

$$n = |\bigcup_{v \in S} N_G(v)| \leq \sum_{v \in S} |N_G(v)| \leq |S| \cdot \Delta(G),$$

or, equivalently, $\gamma_t(G) = |S| \geq n/\Delta(G)$. □

If G is a connected graph of order $n \geq 2$ with $\Delta(G) = n-1$, then a vertex of maximum degree and an arbitrary neighbor of such a vertex form a TD-set in G, and so $\gamma_t(G) = 2 = n - \Delta(G) + 1$ in this trivial case. In the more interesting case when $\Delta(G) \leq n-2$, Cockayne, Dawes, and Hedetniemi [39] established the following upper bound of the total domination number of a graph in terms of the order and maximum degree of the graph.

Theorem 2.12 ([39]). *If G is a connected graph of order $n \geq 3$ and $\Delta(G) \leq n-2$, then $\gamma_t(G) \leq n - \Delta(G)$.*

Haynes and Markus [100] established the following property of graphs that achieve equality in the upper bound of Theorem 2.12. For this purpose, for $1 \leq k \leq n$ they define the *generalized maximum degree* of a graph G, denoted by $\Delta_k(G)$, to be $\max\{|N(S)| : S \subseteq V \text{ and } |S| = k\}$ and noted that $\Delta_1(G) = \Delta(G)$.

Theorem 2.13 ([100]). *Let G be a connected graph of order $n \geq 3$ with $\Delta = \Delta(G) \leq n-2$. Then, $\gamma_t(G) = n - \Delta$ if and only if $\Delta_k(G) = \Delta + k$ for all $k \in \{2, \ldots, \gamma_t(G)\}$.*

We remark that one direction of Theorem 2.13 can be slightly strengthened as follows.

Theorem 2.14. *Let G be a connected graph of order $n \geq 3$ with $\Delta = \Delta(G) \leq n-2$. If $\Delta_{n-\Delta-1}(G) = \Delta + (n - \Delta - 1) = n-1$, then $\gamma_t(G) = n - \Delta$.*

Proof. If $\Delta_{n-\Delta-1}(G) = n-1$, then $|N(S)| < n$ for all $S \subseteq V(G)$, with $|S| = n - \Delta - 1$, implying that $\gamma_t(G) \geq n - \Delta$. By Theorem 2.12, $\gamma_t(G) \leq n - \Delta$. Consequently, $\gamma_t(G) = n - \Delta$. □

A constructive characterization of connected triangle-free graphs G that achieve equality in Theorem 2.12 can be found in [42].

2.3.4 Bounds in Terms of Radius and Diameter

In this section, we present some bounds relating the total domination number in a connected graph with its radius or diameter. DeLaViña et al. [44] showed that the radius provides a lower bound for the total domination number.

Theorem 2.15 ([44]). *If G is a connected graph of order at least two, then $\gamma_t(G) \geq \mathrm{rad}(G)$.*

The following characterization of the case of equality for Theorem 2.15 is given in [44].

Theorem 2.16 ([44]). *Let G be a connected graph of order at least two and let S be a $\gamma_t(G)$-set. Then, $\gamma_t(G) = \mathrm{rad}(G)$ if and only if $G[S]$ has size $\mathrm{rad}(G)/2$.*

Note that if $n \equiv 0 \pmod 4$ and $G = P_n$ (a path of order n), then $\mathrm{rad}(G) = \gamma_t(G) = n/2$, implying that the bound in Theorem 2.16 is sharp for paths of order congruent to zero modulo four. Since $\mathrm{diam}(G) \le 2\mathrm{rad}(G)$ for all connected graphs G, Theorem 2.15 implies that $\gamma_t(G) \ge \mathrm{diam}(G)/2$. One can in fact do slightly better.

Theorem 2.17 ([44]). *If G is a connected graph of order at least two, then $\gamma_t(G) \ge (\mathrm{diam}(G)+1)/2$.*

The following stronger result is established in [138].

Theorem 2.18 ([138]). *If $G = (V,E)$ is a connected graph and $x_1, x_2, x_3 \in V$, then*

$$\gamma_t(G) \ge \frac{1}{4}\left(d(x_1,x_2) + d(x_1,x_3) + d(x_2,x_3)\right).$$

Furthermore if $\gamma_t(G) = \frac{1}{4}(d(x_1,x_2) + d(x_1,x_3) + d(x_2,x_3))$, then the multiset $\{d(x_1,x_2), d(x_1,x_3), d(x_2,x_3)\}$ is equal to $\{2,3,3\}$ modulo four.

To illustrate the sharpness of the bound in Theorem 2.18, take, for example, the graph G to be the double star (a tree with exactly two vertices that are not leaves) on five vertices. Let x_1 and x_2 be two leaves with a common neighbor and let x_3 be a leaf at distance 3 from x_1 in G. Then, $\gamma_t(G) = 2$ and $d(x_1,x_2) + d(x_1,x_3) + d(x_2,x_3) = 8$, implying that $\gamma_t(G) = \frac{1}{4}(d(x_1,x_2) + d(x_1,x_3) + d(x_2,x_3))$.

We remark that Theorem 2.17 is an immediate consequence of Theorem 2.18, due to the following argument. Let G be a connected graph and let x_1 and x_3 be two vertices in G with $d(x_1,x_3) = \mathrm{diam}(G)$. Applying Theorem 2.18 with $x_1 = x_2$ and noting that $d(x_1,x_2) = 0 \notin \{2,3\}$ modulo four, we have that $\gamma_t(G) > (d(x_1,x_2) + d(x_1,x_3) + d(x_2,x_3))/4 = \mathrm{diam}(G)/2$, implying that $\gamma_t(G) \ge (\mathrm{diam}(G)+1)/2$.

The result of Theorem 2.18 is extended to four vertices in [138].

Theorem 2.19 ([138]). *If $G = (V,E)$ is a connected graph and $x_1, x_2, x_3, x_4 \in V$, then $\gamma_t(G) \ge \frac{1}{8}(d(x_1,x_2) + d(x_1,x_3) + d(x_1,x_4) + d(x_2,x_3) + d(x_2,x_4) + d(x_3,x_4))$, and this result is best possible.*

We pose a more general problem in Sect. 18.13 where five or more vertices are considered.

The *center* of a graph G, denoted by $C(G)$, is the set of all vertices of minimum eccentricity. Since every vertex in $C(G)$ is at distance at most $\mathrm{rad}(G)$ from every other vertex, we note that $\mathrm{ecc}(C(G)) \le \mathrm{rad}(G)$. When $\mathrm{ecc}(C(G)) = \mathrm{rad}(G)$, the following theorem provides a slight improvement on Theorem 2.15.

Theorem 2.20 ([44]). *If G is a connected graph of order at least two, then $\gamma_t(G) \ge \mathrm{ecc}(C(G)) + 1$.*

The *periphery* of a graph G, denoted by $B(G)$, is the set of all vertices of maximum eccentricity. A lower bound on the total domination number of a graph in terms of the eccentricity of its periphery, $\mathrm{ecc}(B(G))$, is given in [138].

Theorem 2.21 ([138]). *If G is a connected graph of order at least two, then $\gamma_t(G) \ge (3\mathrm{ecc}(B)+2)/4$.*

2.3.5 *Bounds in Terms of Girth*

The girth of a graph can be used to provide both lower and upper bounds for the total domination number as the following two results illustrate.

Theorem 2.22 ([44]). *If G is a graph of girth g, then $\gamma_t(G) \ge g/2$.*

Theorem 2.23 ([137]). *If G is a connected graph of order n, girth $g \ge 3$, and with $\delta(G) \ge 2$, then*

$$\gamma_t(G) \le \frac{n}{2} + \max\left(1, \frac{n}{2(g+1)}\right),$$

and this bound is sharp.

As $n \ge g$, we have that $(\frac{1}{2} + \frac{1}{g})n \ge \frac{n}{2} + 1$, and so as an immediate consequence of Theorem 2.23, we have the following weaker result.

Theorem 2.24 ([132]). *If G is a graph of order n, minimum degree at least two, and girth $g \ge 3$, then*

$$\gamma_t(G) \le \left(\frac{1}{2} + \frac{1}{g}\right)n.$$

Note that if $n \equiv 2 \pmod 4$ and $G = C_n$, then G has order n, girth $g = n$, and $\gamma_t(G) = (n+2)/2 = \left(\frac{1}{2} + \frac{1}{g}\right)n$, implying that the bound in Theorem 2.24, and therefore also of Theorem 2.23, is sharp for cycles of length congruent to two modulo four. A more detailed discussion of bounds on the total domination number of a graph in terms of its girth is given in Chap. 7. In particular, the sharpness of the bound in Theorem 2.23 is discussed in Sect. 7.3.

2.4 Bounds in Terms of the Domination Number

Every TD-set in a graph is also a dominating set in the graph, implying that $\gamma(G) \le \gamma_t(G)$ for all graphs G with no isolated vertex. Furthermore if S is a $\gamma(G)$-set in a graph G and X is obtained by picking an arbitrary neighbor of each vertex in S, then $|X| \le |S|$ and the set $X \cup S$ is a TD-set in G. Hence, $\gamma_t(G) \le |X| + |S| \le 2|S| = 2\gamma(G)$. This implies the following relationship between the domination and total domination numbers of a graph with no isolated vertex first observed by Bollobás and Cockayne [16].

Theorem 2.25 ([16]). *For every graph G with no isolated vertex, $\gamma(G) \le \gamma_t(G) \le 2\gamma(G)$.*

A more detailed discussion of bounds on the total domination number of a graph in terms of its domination number is given in Sects. 4.6 and 4.7.

Chapter 3
Complexity and Algorithmic Results

3.1 Introduction

In this chapter we discuss complexity and algorithmic results on total domination in graphs. This area, although well studied, is still not developed to the same level as for domination in graphs. We will outline a few of the best-known algorithms and state what is currently known in this field.

3.2 Complexity

The basic complexity question concerning the decision problem for the total domination number takes the following form:

Total Dominating Set
Instance: A graph $G = (V,E)$ and a positive integer k
Question: Does G have a TD-set of cardinality at most k?

For a definition of the graph classes mentioned in this chapter, we refer the reader to the excellent survey on graph classes by Brandstädt, Le, and Spinrad [18] which is regarded as a definitive encyclopedia for the literature on graph classes.

Let graph class $\mathscr{G}_1$ be a proper subclass of a graph class $\mathscr{G}_2$, i.e., $\mathscr{G}_1 \subset \mathscr{G}_2$. If problem $\mathscr{P}$ is NP-complete when restricted to $\mathscr{G}_1$, then it is NP-complete on $\mathscr{G}_2$. Furthermore, any polynomial time algorithm that solves a problem $\mathscr{P}$ on $\mathscr{G}_2$ also solves $\mathscr{P}$ on $\mathscr{G}_1$. Hence it is useful to know the containment relations between certain graph classes. In particular, we have the containment relations described above Table 3.1:

Table 3.1 summarizes the NP-completeness results for the total domination number with the corresponding citation. We abbreviate "NP-complete" by "**NP-c**" and "polynomial time solvable" by "**P**."

M.A. Henning and A. Yeo, *Total Domination in Graphs*, Springer Monographs in Mathematics, DOI 10.1007/978-1-4614-6525-6_3,

$$\begin{aligned}
\text{Bipartite} &\subset \text{comparability} \\
\text{Split} &\subset \text{chordal} \\
\text{Claw-free} &\subset \text{line} \\
\text{Strongly chordal} &\subset \text{dually chordal} \\
\text{Permutation} &\subset k\text{-polygon} \subset \text{circle} \\
\text{Interval} &\subset \text{strongly chordal} \subset \text{chordal} \\
\text{Permutation} &\subset \text{cocomparability} \subset \text{asteroidal triple-free}
\end{aligned}$$

Table 3.1 NP-complete results for the total domination number

Graph Class	NP-completeness result	Citation
General graph	**NP-c**	[170]
Bipartite graph	**NP-c**	[170]
Comparability graph	**NP-c**	[170]
Split graph	**NP-c**	[158]
Chordal graph	**NP-c**	[159]
Line graph	**NP-c**	[165]
line graph of bipartite graph	**NP-c**	[165]
Claw-free graph	**NP-c**	[165]
Circle graph	**NP-c**	[151]
Planar graph, max-degree 3	**NP-c**	[71]
Chordal bipartite graph	**P**	[155, 174]
Interval graph	**P**	[10, 11, 25, 150, 176]
Permutation graph	**P**	[17, 38, 155]
Strongly chordal graph	**P**	[24]
Dually chordal graph	**P**	[155]
Cocomparability graph	**P**	[154, 155]
Asteroidal triple-free graphs	**P**	[156]
DDP-graphs	**P**	[156]
Distance hereditary graph	**P**	[26, 155]
k-polygon graph (fixed $k \geq 3$)	**P**	[155]
Partial k-tree (fixed $k \geq 3$)	**P**	[8, 195]
Trapezoid graph	**P**	[156]

As far as we are aware the class of chordal bipartite graphs is the only class where the complexities of finding a minimum dominating set and a minimum TD-set vary. The problem of finding a minimum dominating set in chordal bipartite graphs is NP-hard, by a result in [41].

3.2.1 Time Complexities

In [155] Kratsch and Stewart give a transformation that implies that any algorithm for domination can be used for total domination for a wide variety of graph classes, such as permutation graphs, dually chordal graphs, and k-polygon graphs. In [155] it is pointed out that this gives an $O(nm^2)$ algorithm for computing a minimum cardinality TD-set in cocomparability graphs, improving on an $O(n^6)$ algorithm in [154]. It also implies an $O(n+m)$ and an $O(n\ln(\ln(n)))$ algorithm for permutation graphs.

In [156] Kratsch gives an $O(n^6)$ algorithm for total domination in asteroidal triple-free graphs. Asteroidal triple-free graphs form a large class of graphs containing interval, permutation, trapezoid, and cocomparability graphs, and therefore, there exist $O(n^6)$ algorithms for these graph classes as well. In fact, in [156] it is shown that there is a $O(n^7)$ algorithm for total domination for the larger class of DDP-graphs. A DDP-graph is a graph where every component has a dominating diametral path, which is a path whose length is equal to the diameter and whose vertex set dominates the graph. Any asteroidal triple-free graph is also a DDP-graph.

In [174] Pradhan gives an $O(n+m)$ algorithm for finding a minimum TD-set in chordal bipartite graphs. In [26] a linear algorithm is also given for distance hereditary graphs.

Bertossi and Gori [11] constructed an $O(n\ln n)$ algorithm for the total domination number of an interval graph of order n.

Adhar and Peng [1] presented efficient parallel algorithms for total domination in interval graphs, while Bertossi and Moretti [12] and Rao and Pandu Rangan [177] presented efficient parallel algorithms for total domination on circular-arc graphs.

In Sect. 3.5 we give a linear time algorithm for finding a minimum TD-set in trees.

3.3 Fixed Parameter Tractability

As determining the total domination number of a graph is NP-hard for most classes of graphs, we need to consider either non-polynomial algorithms or approximation algorithms or heuristics. Towards the end of the last millennium Downey and Fellows [58] introduced a new concept to handle NP-hardness, namely, fixed parameter tractability, which is often abbreviated to FPT.

A problem, $\mathscr{P}$, with size n and a parameter k is fixed parameter tractable (FPT) if it can be solved in time $O(f(k)n^c)$, for some function $f(k)$ not depending on n and some constant c not depending on n or k. For total domination the parameter used is normally the size of the solution. So the question is if there exists an algorithm with complexity $O(f(\gamma_t(G))n^c)$ to determine the total domination number of a graph G. Unfortunately, according to the theory of FPT, it is very unlikely that determining the total domination number is fixed parameter tractable for general graphs. In fact

Table 3.2 FPT-complexity results for the total domination number

Graph class	FPT complexity	Citation
General graphs	$W[2]$-hard	[58]
Graphs with girth 3 or 4*	$W[2]$-hard	[171] (Philip, G.[a])
Girth at least 5*	FPT	(Philip, G.[b])
Planar graph	FPT	[2,58]
Bounded treewidth*	FPT	[2]
d-degenerate graphs*	FPT	[5] (Alon, N.[c])
Bounded maximum degree*	FPT	

*We give some explanation below.
[a]personal communication.
[b]personal communication.
[c]personal communication.

it is shown that the problem is $W[2]$-hard, which according to FPT theory means that the problem is not FPT unless a very unlikely collapse of complexity classes occurs. We will not go into more depth in this area and refer the interested reader to [58] for more information about FPT.

In Table 3.2 we list for which classes of graphs the problem of determining the total domination number is FPT and for which classes it is unlikely to be FPT, by being $W[2]$-hard.

We next describe some of the graph classes listed in Table 3.2 and approaches to determine their FPT complexity.

Girth 3 or 4: In this case the same reductions that were used for connected domination in [171] can be used (Philip, G., personal communication).

Girth at Least Five: Let G be a graph with girth at least 5 and suppose we want to decide if $\gamma_t(G) \le k$, for some k. Let $x \in V(G)$ be arbitrary and note that as the girth is at least five, the vertex x is the only vertex in G that totally dominates more than one vertex in $N_G(x)$. Therefore if $d_G(x) > k$ and $\gamma_t(G) \le k$, then the vertex x must belong to every minimum TD-set in G. If $d_G(x) \le k$, then every TD-set has to include at least one vertex from $N_G(x)$ (in order to totally dominate x), and so by trying all possibilities, we obtain a search tree with branching factor at most k and depth at most k, implying that we can find all TD-sets in G of size at most k in time $O(k^k(n+m))$. Therefore we can also decide if $\gamma_t(G) \le k$ in the same time, implying that the problem is FPT.

Bounded Treewidth: The notion of treewidth was introduced by Robertson and Seymour [178] and plays an important role in their fundamental work on graph minors. We therefore give a brief description of this important notion. A tree decomposition of a graph $G = (V,E)$ is a pair $(\mathscr{Y},T)$ where T is a tree with $V(T) = \{1,2,\ldots,r\}$ (for some $r \ge 1$) and $\mathscr{Y} = \{Y_i \subseteq V(G) \mid i = 1,2,\ldots,r\}$ is a family of subsets of $V(G)$ such that the following holds: (1) $\cup_{i=1}^{r} Y_i = V(G)$; (2) for each edge $e = \{u,v\} \in E(G)$, there exists a Y_i, such that $\{u,v\} \subseteq Y_i$; and (3) for all $v \in V(G)$, the set of vertices $\{i \mid v \in Y_i\}$ forms a connected subtree of T.

Given a tree decomposition, which always exists as we could let $r = 1$ and $Y_1 = V(G)$, the width is $\max\{|Y_i| : i = 1, 2, \ldots, r\} - 1$, and the treewidth of a graph is the smallest possible width that can be obtained by a tree decomposition. A graph is said to have bounded treewidth if this value is bounded by a constant. For example, if G is a tree its treewidth can be shown to be 1.

d-Degenerate Graphs: A graph is called d-degenerate if every induced subgraph has a vertex of degree at most d. It was shown in [5] that it is FPT to find a dominating set of size at most k (the parameter) in a d-degenerate graph. The approach in [5] can also be used to show that the equivalent problem for total domination is FPT (Alon, N., personal communication).

Bounded Maximum Degree: If the maximum degree of a graph is bounded by a constant d, then the graph is d-degenerate and therefore the problem is FPT by the equivalent result for d-degenerate graphs. Alternatively it is also easy to see that the problem is FPT by using a simple search tree. As the size of all neighborhoods is at most d, the branching factor of the search tree is at most d. Further the depth is at most k, giving us the desired FPT algorithm (as d is considered a constant).

3.4 Approximation Algorithms

Let $H_i = 1 + 1/2 + 1/3 + \cdots + 1/i$, which is the ith harmonic number. It is known that $\ln(i) + ec < H_i < \ln(i) + 1/(2i) + ec$, where $ec = 0.57721\ldots$ is the Euler constant. Using a result on the problem *Minimum Set Cover* in [59], the following is shown in [34].

Theorem 3.1 ([34]). *There exists a $(H_{\Delta(G)} - \frac{1}{2})$-approximation algorithm for the problem TOTAL DOMINATION.*

Theorem 3.2 ([34]). *There exists constants $c > 0$ and $D \geq 3$ such that for every $\Delta \geq D$ it is NP-hard to approximate the problem TOTAL DOMINATION within a factor* $\ln(\Delta) - c \ln\ln(\Delta)$ *for bipartite graphs with maximum degree Δ.*

Note that by Theorem 3.2 it is NP-hard to approximate the problem TOTAL DOMINATION within a factor $\ln(\Delta) - c \ln\ln(\Delta)$ for general graphs with maximum degree $\Delta \geq 3$.

3.5 A Tree Algorithm

Laskar, Pfaff, Hedetniemi, and Hedetniemi [159] constructed the first linear algorithm for computing the total domination number of a tree. For ease of presentation, we consider rooted trees. We commonly draw the root of a rooted tree at the top with the remaining vertices at the appropriate level below the root depending

Fig. 3.1 A rooted tree T with its parent array

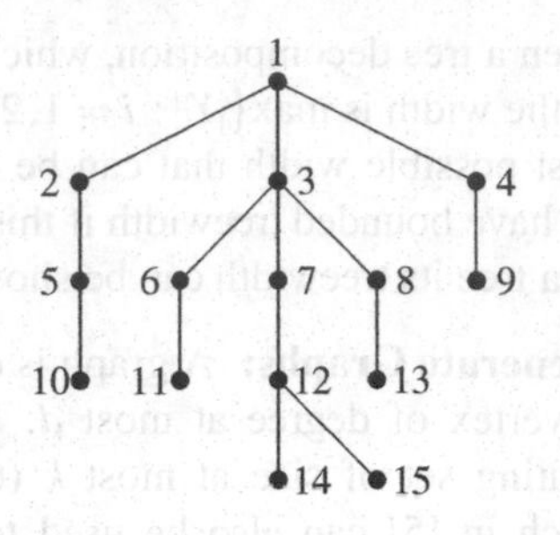

	1	2	3	4	5	6	7	8	9	10	11	12	13	14	15
parent	[0	1	1	1	2	3	3	3	4	5	6	7	8	12	12]

on their distance from the root. The linear algorithm we present here is similar to an algorithm due to Mitchell, Cockayne, and Hedetniemi [167] for computing the domination number of an arbitrary tree in the sense that we root the tree and systematically consider the vertices of the tree, starting from the vertices at furthest distance from the root, and carefully select a TD-set S in such a way that the sum of the distances from the root to the vertices in S is minimum. Given a rooted tree T with root vertex labeled 1 and with $V(T) = \{1,2,\ldots,n\}$, we represent T by a data structure called a *parent array* in which the parent of a vertex labeled i is given by parent$[i]$ with parent$[1] = 0$ (to indicate that the vertex labeled 1 has no parent). We assume that the vertices of T are labeled $1,2,\ldots,n$ so that for $i < j$, vertex i is at level less than or equal to that of vertex j (i.e., $d(1,i) \le d(1,j)$). Figure 3.1 shows an example of a rooted tree T with its parent array.

In Algorithm TREE TOTAL DOMINATION that follows, we will for each vertex, i, keep track of two boolean values In_TD_Set$[i]$ and TD_by_child$[i]$, which both initially will be put to false. As the algorithm progresses some of these values may change to true. Upon completion of the algorithm, all vertices, i, with In_TD_Set$[i] =$ true will be added to the TD-set and all vertices with TD_by_child$[i] =$ true will be totally dominated by one of their children. Below we define vertex 0 to be a *dummy vertex* that does not exist and parent$[1] =$ parent$[0] = 0$. We will never change In_TD_Set$[0]$ but may change TD_by_child$[0]$.

Algorithm TREE TOTAL DOMINATION:

Input: A rooted tree $T = (V,E)$ with $V = \{1,2,\ldots,n\}$ rooted at 1 (where larger values are further from the root) and represented by an array parent$[1\ldots n]$
Output: An array In_TD_Set[] that indicates which vertices belong to the TD-set
Code:

1. **for** $i = 1$ **to** n **do** {
2. TD_by_child$[i]$ = *false*
3. In_TD_Set$[i]$ = *false* }
4. ***for*** $i = n$ ***to*** 2 ***do***
5. ***if*** (TD_by_child$[i]$ = *false)* ***and*** (In_TD_Set[parent$[i]$] = *false)* ***then*** {

6. In_TD_Set[parent[i]] $=$ *true*
7. TD_by_child[parent[parent[i]]] $=$ *true* }
8. ***if*** (TD_by_child[1] $=$ *false*) ***then***
9. In_TD_Set[2] $=$ *true*

We will now outline why the Algorithm TREE TOTAL DOMINATION does produce a minimum TD-set. Let $S(i)$ denote all vertices, j, with In_TD_Set[j] $=$ true, just after performing line 7 with the value i. We will show that property $P(i)$ below holds for all $i = n, n-1, n-2, \dots, 2$, by induction.

Property $P(i)$**:** $S(i)$ totally dominates $\{i, i+1, i+2, \dots, n\}$ and $S(i)$ is a subset of some minimum TD-set in G. Furthermore the array TD_by_child is updated correctly.

Property $P(n)$ is clearly true as the set $S(n)$ only contains the unique neighbor of vertex n. So let $2 \le i < n$ and assume that property $P(i+1)$ holds and that $S(i+1) \subseteq Q_{i+1}$, where Q_{i+1} is a minimum TD-set in G. We will now consider the two possible outcomes of line 5.

If TD_by_child[i] $=$ false and In_TD_Set[parent[i]] $=$ false, then some vertex, t, in Q_{i+1} must totally dominate i. If $t =$ parent[i], then $S(i) \subseteq Q_{i+1}$, and therefore $S(i)$ is a subset of an optimal solution. Moreover as $S(i+1)$ totally dominates $\{i+1, i+2, \dots, n\}$ and the vertex parent[i] $\in S(i)$ totally dominates the vertex i, the set $S(i)$ totally dominates $\{i, i+1, i+2, \dots, n\}$. Hence we may assume that t is a child of i, for otherwise the desired result follows. Let $Q_i = (Q_{i+1} \setminus \{t\}) \cup \{\text{parent}[i]\}$ and note that Q_i is a minimum TD-set of G as all neighbors of t, except for the vertex i, are totally dominated by $S(i+1)$ and $S(i+1) \subseteq Q_i$. Further, $S(i) = S(i+1) \cup \{\text{parent}[i]\} \subseteq Q_i$, and so the set $S(i)$ is a subset of an optimal solution. As before since the set $S(i+1)$ totally dominates $\{i+1, i+2, \dots, n\}$ and the vertex parent[i] $\in S(i)$ totally dominates the vertex i, the set $S(i)$ totally dominates $\{i, i+1, i+2, \dots, n\}$.

If TD_by_child[i] $=$ true or In_TD_Set[parent[i]] $=$ true, then the vertex i is totally dominated by $S(i+1)$ and therefore $S(i)$ totally dominates $\{i, i+1, i+2, \dots, n\}$. Furthermore $S(i) = S(i+1) \subseteq Q_{i+1}$ which implies that $S(i)$ is a subset of an optimal solution.

It remains for us to verify that the array TD_by_child is updated correctly. If $j \in S(i)$, then let i_j be the value of i when In_TD_Set[j] is put to true. That is, parent[i_j] $= j$. In line 7, TD_by_child[parent[j]] is then put to true. Furthermore suppose that TD_by_child[j'] $=$ true for some vertex j'. Let i'_j be the value of i when TD_by_child[j'] was put to true. That is, parent[parent[i'_j]] $= j'$. In line 6, parent[i'_j] is then put to true, which implies that the child parent[i'_j] of the vertex j' belongs to the TD-set, as desired. This implies that TD_by_child is updated correctly. Therefore, Property $P(i)$ holds.

As Property $P(2)$ holds, we note that the set $S(2)$ is a subset of an optimal solution Q_2, say, and $S(2)$ totally dominates all vertices in G, except possibly for vertex 1. Therefore if vertex 1 is totally dominated by $S(2)$, then $S(2) = Q_2$ and S_2 is an optimal solution. On the other hand, if vertex 1 is not totally dominated by $S(2)$, then some vertex, z, in Q_2 must totally dominate vertex 1. Thus, z is a child of vertex 1 and the set $S(1) = Q_1 = (Q_2 \setminus \{z\}) \cup \{2\}$ is an optimal solution. Therefore TREE TOTAL DOMINATION does indeed find an optimal solution.

It is also not difficult to see that the time complexity of the algorithm is $O(n)$ for trees of order n.

When we run our algorithm on the rooted tree T in Fig. 3.1, the following steps will be performed, yielding the set $\{12,8,7,6,5,4,3,2,1\}$ as a minimum TD-set in T, as the if statement in line 8 will be false.

i	Action	i	Action	i	Action
15	In_TD_Set[12] = true	10	In_TD_Set[5] = true	5	In_TD_Set[2] = true
14	None	9	In_TD_Set[4] = true	4	In_TD_Set[1] = true
13	In_TD_Set[8] = true	8	In_TD_Set[3] = true	3	None
12	In_TD_Set[7] = true	7	None	2	None
11	In_TD_Set[6] = true	6	None		

3.6 A Simple Heuristic

In this section we will give a simple heuristic that finds a TD-set in a graph G of size at most $n(G)(1+\ln\delta(G))/\delta(G)$. Furthermore, in Theorem 3.4 it is shown that there exist graphs where there do not exist TD-sets of size much smaller than this.

The heuristic simply keeps choosing the vertex that totally dominates the maximum number of vertices which are not yet totally dominated and adding this vertex to our TD-set. It continues to do this until all vertices are totally dominated. Below is the code for this heuristic, where T will be our TD-set and Not_TD will contain the vertices that are not totally dominated by T.

Heuristic TOTAL DOMINATION:

Input: A graph $G = (V,E)$ with minimum degree $\delta \geq 1$ and order n.
Output: A TD-set T of G.
Code:

1. $T = \emptyset$
2. Not_TD $= V(G)$
3. ***While*** Not_TD $\neq \emptyset$ ***do*** {
4.   ***Let*** $x \in V(G)$ ***have maximum*** $d_{\text{Not\_TD}}(x)$
5. $T = T \cup \{x\}$
6.   Not_TD = Not_TD $\setminus N(x)$ }
7. }

The fact that Heuristic TOTAL DOMINATION finds a TD-set of size at most $n(G)(1+\ln\delta(G))/\delta(G)$ was shown in [130], by considering the open neighborhood hypergraph, H_G, of G. We will below give a proof of this result not involving hypergraphs.

It is not too difficult to show that with the correct implementation, the Heuristic TOTAL DOMINATION can be made to run in time $O(n+m)$. It would require us to keep a data structure (such as a hash table) that allows us to perform line 4 in constant time. We would then need to continuously update the degrees $d_{\text{Not\_TD}}()$ and note that we can decrease such degrees at most $2m(G)$ times (when removing vertices from Not_TD as $\sum_{v\in G} d_G(v) = 2m(G)$). We will leave the details of the complexity computations to the interested reader.

In fact if all we want is the bound $n(G)(1+\ln\delta(G))/\delta(G)$, then we can find a TD-set of size at most in this time $O(n+\delta(G)n)$ by considering the open neighborhood hypergraph, H_G, of G and making it $\delta(G)$-uniform. However in practice the heuristic given in this section will generally perform better (when G is not regular).

Theorem 3.3. *If G is a graph with minimum degree $\delta \geq 1$, order n, and size m, then Heuristic TOTAL DOMINATION produces a TD-set T in G satisfying*

$$|T| \leq \left(\frac{1+\ln\delta}{\delta}\right)n.$$

The complexity of the algorithm is $O(n+m)$.

Proof. Clearly the theorem is true for $\delta = 1$, so assume that $\delta \geq 2$. Let T_δ be the vertices added to T in line 5 of the Heuristic TOTAL DOMINATION with $d_{\text{Not\_TD}}(x) \geq \delta$ (see line 4 of the heuristic). Let T_i be the vertices added to T with $d_{\text{Not\_TD}}(x) = i$ for $i = 1,2,\ldots,\delta-1$. Let Not_TD$_i$ denote the set Not_TD after having added all vertices in $T_{i+1} \cup T_{i+2} \cup \cdots \cup T_\delta$ to T. That is, $\text{Not\_TD}_i = V(G) \setminus (\cup_{j=i+1}^{\delta} N(T_j))$. Furthermore let $t_i = |T_i|$ for all $i = 1,2,\ldots,\delta$. Then,

$$T = \bigcup_{i=1}^{\delta} T_i \quad \text{and} \quad |T| = \sum_{i=1}^{\delta} t_i.$$

Given any $i \in \{1,2,\ldots,\delta\}$, we have that

$$\sum_{v\in \text{Not\_TD}_i} d_G(v) = \sum_{w\in V(G)} d_{\text{Not\_TD}_i}(w), \tag{3.1}$$

since every edge with exactly one end in Not_TD$_i$ adds 1 to both sides of the equation, every edge with both ends in Not_TD$_i$ adds 2 to both sides of the equation, and all other edges do not affect either side. Further since $d_{\text{Not\_TD}_i}(w) \leq i$ for all $w \in V(G)$, Eq. (3.1) implies that

$$\sum_{v \in \text{Not\_TD}_i} d_G(v) \leq in. \tag{3.2}$$

For $1 \leq j \leq i \leq \delta - 1$, every vertex $x \in T_j$ removes j vertices from Not_TD_j, while every vertex $x \in T_\delta$ removes at least δ vertices from Not_TD_δ. Hence, for each value of i with $i \in \{1,2,\ldots,\delta\}$, we have that

$$|\text{Not\_TD}_i| = \sum_{j=1}^{i} \left(\sum_{x \in T_j} d_{\text{Not\_TD}_j}(x) \right) \geq \sum_{j=1}^{i} jt_j. \tag{3.3}$$

By Eqs. (3.2) and (3.3), we therefore have that for each value of i with $i \in \{1,2,\ldots,\delta\}$,

$$\delta \sum_{j=1}^{i} jt_j \leq \delta |\text{Not\_TD}_i| \leq \sum_{v \in \text{Not\_TD}_i} d_G(v) \leq in,$$

and so,

$$\sum_{j=1}^{i} jt_j \leq \frac{in}{\delta}.$$

Therefore there exist nonnegative real numbers $\varepsilon_0, \varepsilon_1, \ldots, \varepsilon_\delta$ such that

$$\sum_{j=1}^{i} jt_j = \frac{in}{\delta} - \varepsilon_i$$

for all $i = 0, 1, \ldots, \delta$, where $\varepsilon_0 = 0$. This implies that for each value of j with $j \in \{1,2,\ldots,\delta\}$, we have the following equation:

$$jt_j = \sum_{i=1}^{j} it_i - \sum_{i=1}^{j-1} it_i = \frac{jn}{\delta} - \varepsilon_j - \left[\frac{(j-1)n}{\delta} - \varepsilon_{j-1} \right] = \frac{n}{\delta} - (\varepsilon_j - \varepsilon_{j-1}).$$

Thus,

$$|T| = \sum_{i=1}^{\delta} t_i = \sum_{i=1}^{\delta} \left(\frac{n}{i\delta} \right) - \frac{\varepsilon_\delta}{\delta} - \sum_{i=1}^{\delta-1} \varepsilon_i \left(\frac{1}{i} - \frac{1}{i+1} \right) \leq \sum_{i=1}^{\delta} \left(\frac{n}{i\delta} \right).$$

Therefore,

$$|T| \leq \frac{n}{\delta} \sum_{i=1}^{\delta} \frac{1}{i} \leq \frac{n}{\delta}(1 + \ln \delta) = \left(\frac{1 + \ln \delta}{\delta} \right) n.$$

This completes the proof of the bound in the theorem. The time complexity follows from the discussion preceding the statement of the theorem. □

For sufficiently large δ, the bound on the size of the TD-set produced by Heuristic TOTAL DOMINATION is close to optimal, as can be deduced from the following result.

Theorem 3.4 ([207]). *For every $k \geq 1$, there exists a bipartite k-regular graph, G, with $\gamma_t(G) > \left(\frac{0.1 \ln(k)}{k}\right) n(G)$.*

Chapter 4
Total Domination in Trees

4.1 Introduction

In this chapter, we present results on total domination in trees. For a linear algorithm to compute the total domination of a tree, see Sect. 3.5.

4.2 Bounds on the Total Domination Number in Trees

Chellali and Haynes [29, 30] established the following lower and upper bounds on the total domination of a tree in terms of the order, number of leaves, and number of support vertices in the tree.

Theorem 4.1 ([29,30]). *Let T be a tree with $n \geq 3$ vertices, ℓ leaves, and s support vertices. Then,* $(n+2-\ell)/2 \leq \gamma_t(T) \leq (n+s)/2$.

Since the number of support vertices in a tree of order at least 2 is at most the number of leaves in the tree, we have the following immediate consequence of Theorem 4.1.

Theorem 4.2 ([29,30]). *If T is a tree with $n \geq 2$ vertices and ℓ leaves, then* $\gamma_t(T) \leq (n+\ell)/2$.

A characterization of the trees achieving equality in the upper bound of Theorem 4.2 can readily be deduced from a result due to Chen and Sohn [33]. In order to state their result, we need a variation of total domination, called locating–total domination, which we discuss later in Sect. 17.4. A *locating–total dominating set*, abbreviated LTD-set, in a graph $G = (V,E)$ is a TD-set S with the property that distinct vertices in $V \setminus S$ are totally dominated by distinct subsets of S. The *locating–total domination number*, denoted $\gamma_t^L(G)$, of G is the minimum cardinality of a

M.A. Henning and A. Yeo, *Total Domination in Graphs*, Springer Monographs in Mathematics, DOI 10.1007/978-1-4614-6525-6_4,

LTD-set of G. Chen and Sohn [33] describe a procedure to build a family Γ of labeled trees as follows. Each vertex v in a tree from the family Γ receives a label, namely, A, B, or C, called its *status* and denoted by sta(v), as follows.

Definition 4.1 ([33]). Let Γ be the family of labeled trees $T = T_k$ that can be obtained as follows. Let T_0 be a path P_6 on six vertices in which the two leaves have status C, the two support vertices have status A, and the two central vertices have status B. If $k \geq 1$, then T_k can be obtained recursively from T_{k-1} by one of the following operations:

- Operation τ_1. For any $y \in V(T_{k-1})$, if $sta(y) = C$ and y is a leaf of T_{k-1}, then add a path $xwvz$ and edge xy. Let $sta(x) = sta(w) = B$, $sta(v) = A$, and $sta(z) = C$.
- Operation τ_2. For any $y \in V(T_{k-1})$, if $sta(y) = B$, then add a path xwv and edge xy. Let $sta(x) = B$, $sta(w) = A$, and $sta(v) = C$.

The labeled tree T_0 and the two operations τ_1 and τ_2 are illustrated in Fig. 4.1.

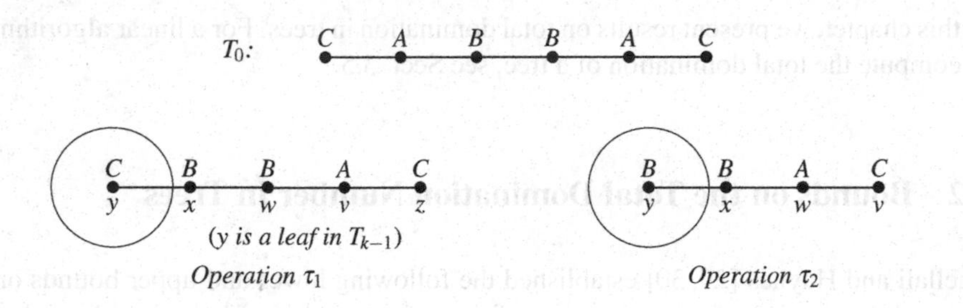

Fig. 4.1 Operations to build Γ

Theorem 4.3 ([33]). *Let T be a tree of order $n \geq 3$ with ℓ leaves. Then the following hold:*
(a) $\gamma_t^L(T) \leq (n + \ell(T))/2$.
(b) *If $T \in \Gamma$, then* $\gamma_t(T) = (n + \ell(T))/2$.
(c) $\gamma_t^L(T) = (n + \ell(T))/2$ *if and only if* $T \in \Gamma$.

The following is observed in [49]. Suppose that T is a tree of order $n \geq 3$ satisfying $\gamma_t(T) = (n + \ell)/2$. Since every LTD-set is also a TD-set, we have that $\gamma_t(T) \leq \gamma_t^L(T)$. By Theorem 4.3(a), $\gamma_t^L(T) \leq (n + \ell(T))/2$, implying that $\gamma_t(T) = \gamma_t^L(T) = (n + \ell(T))/2$. Hence, by Theorem 4.3(c), $T \in \Gamma$. Conversely by Theorem 4.3(b), if $T \in \Gamma$, then $\gamma_t(T) = (n + \ell(T))/2$. Hence we have the following constructive characterization of trees that achieve equality in the upper bound of Theorem 4.2.

Theorem 4.4 ([33]). *For any tree T of order $n \geq 3$, $\gamma_t(T) = (n + \ell(T))/2$ if and only if $T \in \Gamma$.*

4.3 Total Domination and Open Packings

Any TD-set must have a nonempty intersection with every open neighborhood. Hence we observe that if G is a graph with no isolated vertices, then $\gamma_t(G) \geq \rho^o(G)$, where we recall that $\rho^o(G)$ is the maximum cardinality of an open packing in G. Rall [175] was the first to prove equality between the total domination number and the open packing number of any tree of order at least two.

Theorem 4.5 ([175]). *For every tree T of order at least 2, $\gamma_t(T) = \rho^o(T)$.*

In fact one can see that Theorem 4.5 holds by examining the Algorithm TREE TOTAL DOMINATION in Sect. 3.5. In this algorithm we considered every vertex in the tree, starting with vertices furthest from the root and working our way towards the root. If a vertex, v, did not have any children in the TD-set we were building, then we put the parent of v in the TD-set, which we will denote by S. We could also have put the vertex v in a set OP, which had initially been set equal to the empty set. When considering the root itself, we put any neighbor of the root in S if the root was not yet totally dominated, and in this case we would put the root in OP. Once the process is complete we have that $\gamma_t(T) = |S| = |OP|$ and one can show that OP is an open packing. As observed earlier, $\gamma_t(G) \geq \rho^o(G)$ for all graphs G with no isolated vertex. In particular, $\gamma_t(T) \geq \rho^o(T)$. Therefore,

$$|OP| = |S| = \gamma_t(T) \geq \rho^o(T) \geq |OP|. \tag{4.1}$$

Hence we must have equality throughout the above inequality chain (4.1), and so $\gamma_t(T) = \rho^o(T)$.

4.4 Vertices Contained in All, or in No, Minimum TD-Sets

Using Algorithm TREE TOTAL DOMINATION in Sect. 3.5 we can decide if a vertex is in all minimum TD-sets or in no minimum TD-sets or in some, but not all minimum TD-sets. For example, if T is a tree and $v \in V(T)$, then v belongs to no minimum TD-set if and only if $\gamma_t(T') > \gamma_t(T)$, where T' is obtained from T by adding a pendent edge to v. With a little more work it is also possible to decide if v belongs to all minimum TD-set. However a nicer characterization was given in [40].

In order to illustrate the characterization in [40], let the sets $\mathscr{A}_t(G)$ and $\mathscr{N}_t(G)$ of a graph G be defined as follows:

$\mathscr{A}_t(G) = \{v \in V(G) \mid v \text{ is in every } \gamma_t(G)\text{-set}\}$.
$\mathscr{N}_t(G) = \{v \in V(G) \mid v \text{ is in no } \gamma_t(G)\text{-set}\}$.

Let T be a tree. In order to determine if a vertex v belongs to $\mathscr{A}_t(T)$ or $\mathscr{N}_t(T)$ for some specified vertex $v \in V(T)$, we consider T to be rooted at the vertex v. We may assume that v is not a support vertex since otherwise it belongs to all TD-sets.

We denote the set of leaves in T by $L(T)$ and the set of leaves in $T = T_v$ distinct from v by $L(v)$, that is, $L(v) = D(v) \cap L(T)$. For $j = 0,1,2,3$, we define

$$L_T^j(v) = \{u \in L(v) \mid d_T(u,v) \equiv j \pmod 4\}.$$

If the tree T is clear from the context, we simply write $L^j(v)$ rather than $L_T^j(v)$. We next describe a technique called *tree pruning*. The pruning of T is performed with respect to the root, v. If $d(u) \leq 2$ for each $u \in V(T_v) \setminus \{v\}$, then let $\overline{T}_v = T$. Otherwise, let u be a branch vertex (that is, a vertex of degree at least 3) at maximum distance from v; note that $|C(u)| \geq 2$ and $d(x) \leq 2$ for each $x \in D(u)$. We now apply the following pruning process:

- If $|L^2(u)| \geq 1$, then delete $D(u)$ and attach a path of length 2 to u.
- If $|L^1(u)| \geq 1$, $L^2(u) = \emptyset$ and $|L^3(u)| \geq 1$, then delete $D(u)$ and attach a path of length 2 to u.
- If $|L^1(u)| \geq 1$ and $L^2(u) = L^3(u) = \emptyset$, then delete $D(u)$ and attach a path of length 1 to u.
- If $L^1(u) = L^2(u) = \emptyset$ and $|L^3(u)| \geq 1$, then delete $D(u)$ and attach a path of length 3 to u.
- If $L^1(u) = L^2(u) = L^3(u) = \emptyset$, then delete $D(u)$ and attach a path of length 4 to u.

This step of the pruning process, where all the descendants of u are deleted and a path of length 1, 2, 3, or 4 is attached to u to give a tree in which u has degree 2, is called a *pruning of T_v at u*. We repeat the above process until a tree $\overline{T}_v$ is obtained with $d(u) \leq 2$ for each $u \in V(\overline{T}_v) \setminus \{v\}$, and we call the resulting tree $\overline{T}_v$ a *pruning* of T_v. The tree $\overline{T}_v$ is unique. Thus, to simplify notation, we write $\overline{L}^j(v)$ instead of $L^j_{\overline{T}_v}(v)$.

To illustrate the pruning process, consider the tree T in Fig. 4.2. The vertices u and w are branch vertices at maximum distance 2 from v. Since $|L^2(u)| = 1$, we delete $D(u)$ and attach a path of length 2 to u. Since $|L^1(w)| = 2$ and $L^2(w) = L^3(w) = \emptyset$, we delete $D(w)$ and attach a path of length 1 to w. This pruning of T_v at u and w produces the intermediate tree shown in Fig. 4.2. In this tree, the vertices x and y are branch vertices at maximum distance 1 from v. Since $|L^2(x)| = 1$, we delete $D(x)$ and attach a path of length 2 to x. Since $|L^1(y)| = 1$, $L^2(y) = \emptyset$, and $|L^3(y)| = 1$, we delete $D(y)$ and attach a path of length 2 to y. This pruning of T_v at x and y produces the pruning $\mathscr{T}_v$ of T_v.

The following characterization of the sets $\mathscr{A}_t(T)$ and $\mathscr{N}_t(T)$ for an arbitrary tree T is presented in [40].

Theorem 4.6 ([40]). *Let v be a vertex of a tree T. Then:*

(a) $v \in \mathscr{A}_t(T)$ if and only if v is a support vertex or $|\overline{L}^1(v) \cup \overline{L}^2(v)| \geq 2$.
(b) $v \in \mathscr{N}_t(T)$ if and only if $\overline{L}^1(v) \cup \overline{L}^2(v) = \emptyset$.

To illustrate Theorem 4.6, note that in the pruning $\mathscr{T}_v$ of the tree T in Fig. 4.2, $|\overline{L}^0(v)| = |\overline{L}^1(v)| = 0$, $|\overline{L}^2(v)| = 1$, and $|\overline{L}^3(v)| = 3$; that is, $|\overline{L}^1(v) \cup \overline{L}^2(v)| = 1$. Hence, by Theorem 4.6, $v \notin \mathscr{A}_t(T) \cup \mathscr{N}_t(T)$.

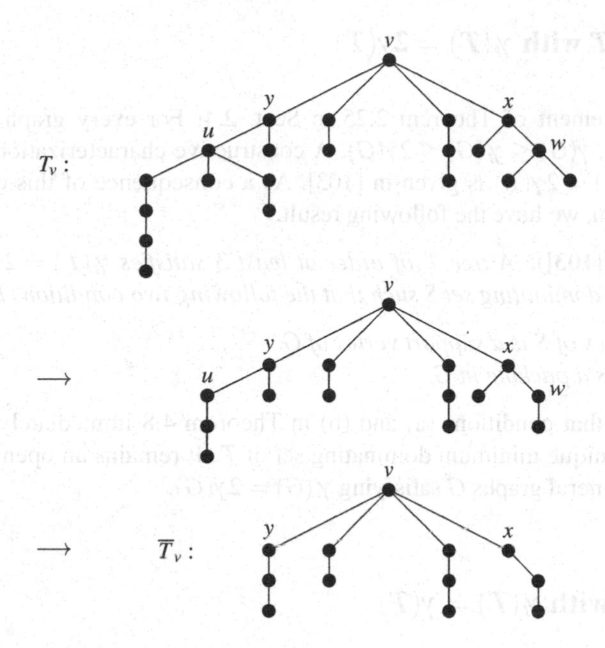

Fig. 4.2 The pruning $\mathscr{T}_v$ of the tree T_v

4.5 Trees with Unique Minimum TD-Sets

In [87] trees having unique minimum TD-sets are investigated. Three equivalent conditions for a tree to have a unique minimum TD-set are provided, and a constructive characterization of such trees is given. Recall that if $G = (V,E)$ is a graph and $S \subseteq V$, then the *open S-private neighborhood* of a vertex $v \in S$ is defined by $\mathrm{pn}(v,S) = \{w \in V\ N_G(w) \cap S = \{v\}\}$.

Theorem 4.7 ([87]). *Let T be a tree of order $n \geq 2$. Then the following conditions are equivalent:*

(a) T has a unique minimum TD-set.
(b) T has a $\gamma_t(T)$-set S for which every vertex in S is a support vertex or satisfies $|\mathrm{pn}(v,S)| \geq 2$.
(c) T has a $\gamma_t(T)$-set S for which $\gamma_t(T-v) > \gamma_t(T)$ for every $v \in S$ that is not a support vertex.
(d) For every vertex $v \in V(T)$, v is a support vertex or $|\overline{L}^1(v) \cup \overline{L}^2(v)| \neq 1$.

4.6 Trees T with $\gamma_t(T) = 2\gamma(T)$

Recall the statement of Theorem 2.25 in Sect. 2.4: For every graph G with no isolated vertex, $\gamma(G) \le \gamma_t(G) \le 2\gamma(G)$. A constructive characterization of trees T satisfying $\gamma_t(T) = 2\gamma(T)$ is given in [103]. As a consequence of this constructive characterization, we have the following result.

Theorem 4.8 ([103]). *A tree T of order at least 3 satisfies $\gamma_t(T) = 2\gamma(T)$ if and only if T has a dominating set S such that the following two conditions hold:*

(a) *Every vertex of S is a support vertex of G.*
(b) *The set S is a packing in G.*

We remark that conditions (a) and (b) in Theorem 4.8 immediately imply that the set S is a unique minimum dominating set of T. It remains an open problem to characterize general graphs G satisfying $\gamma_t(G) = 2\gamma(G)$.

4.7 Trees with $\gamma_t(T) = \gamma(T)$

For any two graph parameters λ and μ, we define a graph G to be a (λ,μ)-graph if $\lambda(G) = \mu(G)$. In [56] it is shown how to generate all (ρ,γ_t)-graphs and a constructive characterization of (ρ,γ_t)-trees is provided, which is particularly useful due to the following theorem.

Theorem 4.9 (Moon and Meir [166]). *For a tree T, $\gamma(T) = \rho(T)$.*

By Theorem 4.9 we note that the constructive characterization of (ρ,γ_t)-trees in [56] is also a characterization of (γ,γ_t)-trees. In fact, by Theorem 4.5, $\gamma_t(T) = \rho^o(T)$ for all trees, T, and therefore any (γ,γ_t)-tree is also a (ρ,ρ^o)-tree. The key to the constructive characterization of (ρ,γ_t)-trees is to find a labeling of the vertices that indicates the roles each vertex plays in the sets associated with both parameters.

We define a (ρ,γ_t)-*labeling* of a graph $G = (V,E)$ as a partition $S = (S_A,S_B,S_C,S_D)$ of V such that $S_A \cup S_D$ is a minimum TD-set, $S_C \cup S_D$ is a maximum packing, and $|S_A| = |S_C|$. The following lemma can easily be proven as if G has a (ρ,γ_t)-labeling, then $\rho(G) = |S_C \cup S_D| = |S_A \cup S_D| = \gamma_t(G)$. Conversely if $\gamma_t(G) = \rho(G)$, then let R be a minimum TD-set and let Q be a maximum packing and create a (ρ,γ_t)-labeling by letting $S_A = R \setminus Q$, $S_B = V(G) \setminus (R \cup Q)$, $S_C = Q \setminus R$, and $S_D = Q \cap R$.

Lemma 4.1 ([56]). *A graph is a (ρ,γ_t)-graph if and only if it has a (ρ,γ_t)-labeling.*

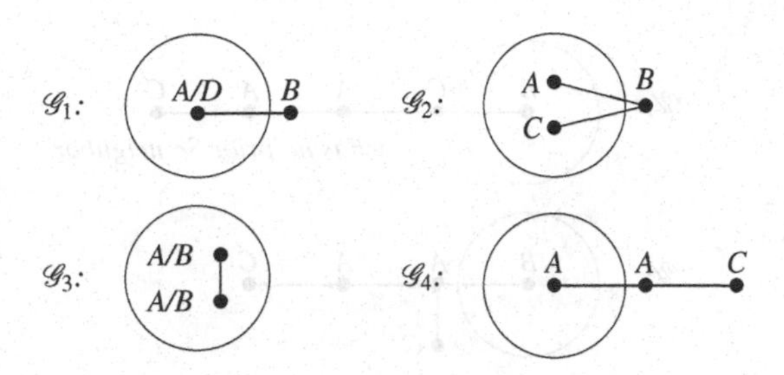

Fig. 4.3 The four $\mathscr{G}_i$ operations

If S is a (ρ,γ_t)-labeling of a (ρ,γ_t)-graph, S, then the *label* or *status* of a vertex v, denoted $\text{sta}(v)$, is the letter $x \in \{A,B,C,D\}$ such that $v \in S_x$. A *labeled* graph is simply one where each vertex is labeled with either A, B, C, or D.

We now define some graph operations:

- *Operation* $\mathscr{G}_1$. Assume $\text{sta}(y) \in \{A,D\}$. Add a vertex x and the edge xy. Let $\text{sta}(x) = B$.
- *Operation* $\mathscr{G}_2$. Assume $\text{sta}(y) = A$ and $\text{sta}(z) = C$. Add a vertex x and the edges xy and xz. Let $\text{sta}(x) = B$.
- *Operation* $\mathscr{G}_3$. Assume $\text{sta}(x), \text{sta}(y) \in \{A,B\}$. Add the edge xy.
- *Operation* $\mathscr{G}_4$. Assume $\text{sta}(y) = A$. Add a path yxw and let $\text{sta}(x) = A$ and $\text{sta}(w) = C$.

These operations are illustrated in Fig. 4.3.

Theorem 4.10 ([56]). *A labeled graph is a (ρ,γ_t)-graph if and only if it can be obtained from a disjoint union of P_4's, with end-vertices labeled C and internal vertices labeled A, using operations $\mathscr{G}_1$ through $\mathscr{G}_4$.*

We now consider (ρ,γ_t)-trees. By Theorem 4.10 we note that the smallest (ρ,γ_t)-tree is the path P_4. It has a unique (ρ,γ_t)-labeling, where the leaves have status C and internal vertices have status A. Now, define the following three operations which are illustrated in Fig. 4.4:

- *Operation* $\mathscr{U}_1$. Take a vertex y of status B which has no neighbor of status C, add a labeled P_4, and join y to a leaf of the P_4.
- *Operation* $\mathscr{U}_2$. Add a labeled P_4, and join a vertex y of status B to an internal vertex of the P_4.
- *Operation* $\mathscr{U}_3$. Add a labeled P_4 and a vertex y' labeled B, and attach to a vertex y of status B or C the added vertex y' and join y' to an internal vertex of the added labeled P_4.

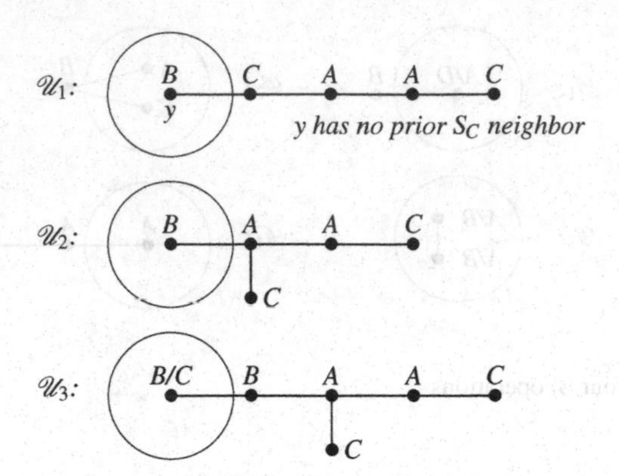

Fig. 4.4 The three $\mathscr{U}_i$ operations

Theorem 4.11 ([56]). *A labeled tree is a (ρ, γ_t)-tree if and only if it can be obtained from a labeled P_4 using the operations $\mathscr{G}_1$, $\mathscr{G}_4$, $\mathscr{U}_1$, $\mathscr{U}_2$, and $\mathscr{U}_3$.*

Chapter 5
Total Domination and Minimum Degree

5.1 Introduction

As remarked in Chap. 3, the decision problem to determine the total domination number of a graph is NP-complete. Hence it is of interest to determine upper bounds on the total domination number of a graph in terms of its minimum degree. In this chapter we consider the following two problems.

Problem 5.1. For a connected graph G with minimum degree $\delta \geq 1$ and order n, find an upper bound $f(\delta, n)$ on $\gamma_t(G)$ in terms of δ and n.

Problem 5.2. Characterize the connected graphs G with minimum degree $\delta \geq 1$ and large-order n satisfying $\gamma_t(G) = f(\delta, n)$.

5.2 A General Bound Involving Minimum Degree

Using probabilistic arguments, Alon [3] proved that for $\delta > 1$, if H is a δ-uniform hypergraph with n vertices and m edges, then $\tau(H) \leq \frac{\ln \delta}{\delta}(n+m)$. In the special case when $n = m$, we therefore have by Theorem 1.2 that if G is a graph with minimum degree $\delta > 1$ and order n and if H is the hypergraph obtained from the *ONH*, H_G, of G by shrinking all edges of H_G, if necessary, to edges of size δ, then $\gamma_t(G) = \tau(H_G) \leq \tau(H) \leq (\frac{2 \ln \delta}{\delta})n$. We can, however, do slightly better by applying Theorem 3.3 stated in Chap. 3 since as an immediate consequence of this theorem, we have the following result.

Theorem 5.1. *If G is a graph with minimum degree $\delta \geq 1$ and order n, then*

$$\gamma_t(G) \leq \left(\frac{1 + \ln \delta}{\delta} \right) n.$$

M.A. Henning and A. Yeo, *Total Domination in Graphs*, Springer Monographs in Mathematics, DOI 10.1007/978-1-4614-6525-6_5,

For sufficiently large δ, the upper bound in Theorem 5.1 is close to optimal, as is shown in Theorem 3.4. Hence for sufficiently large minimum degree δ, Problem 5.1 is essentially solved.

5.3 Minimum Degree One

When $\delta = 1$, we have the results established in Sect. 2.3.1. We restate these results for completeness.

Theorem 5.2 ([37]). *If G is a connected graph of order $n \geq 3$, then $\gamma_t(G) \leq 2n/3$.*

Theorem 5.3 ([20]). *Let G be a connected graph of order $n \geq 3$. Then $\gamma_t(G) = 2n/3$ if and only if G is C_3, C_6 or $H \circ P_2$ for some connected graph H.*

5.4 Minimum Degree Two

If G is a graph of order n that consists of a disjoint union of 3-cycles and 6-cycles, then $\gamma_t(G) = 2n/3$. Hence the upper bound in Theorem 5.2 cannot be improved if we simply restrict the minimum degree to be two. However if we impose the additional restriction that G is connected, then Sun [194] showed that the upper bound in Theorem 5.2 can be improved.

Theorem 5.4 ([194]). *If G is a connected graph of order n with $\delta(G) \geq 2$, then $\gamma_t(G) \leq \lfloor \frac{4}{7}(n+1) \rfloor$.*

The bound of Sun [194] in Theorem 5.4 can be improved slightly if we forbid six graphs of small orders. Let H'_{10} and H_{10} be the two graphs shown in Fig. 5.1a and b, respectively.

Theorem 5.5 ([102]). *If $G \notin \{C_3, C_5, C_6, C_{10}, H_{10}, H'_{10}\}$ is a connected graph of order n with $\delta(G) \geq 2$, then $\gamma_t(G) \leq 4n/7$.*

In order to characterize the connected graphs of large order with total domination number exactly four-sevenths their order, let $\mathscr{F}$ be the family of all graphs that can be obtained from a connected graph F of order at least 2 as follows: For each vertex

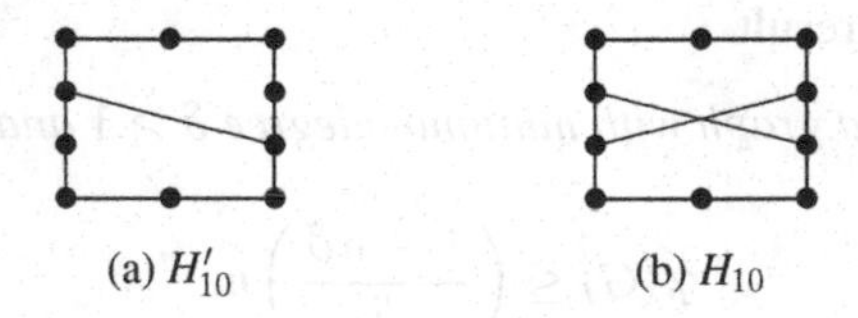

Fig. 5.1 The graphs H'_{10} and H_{10}

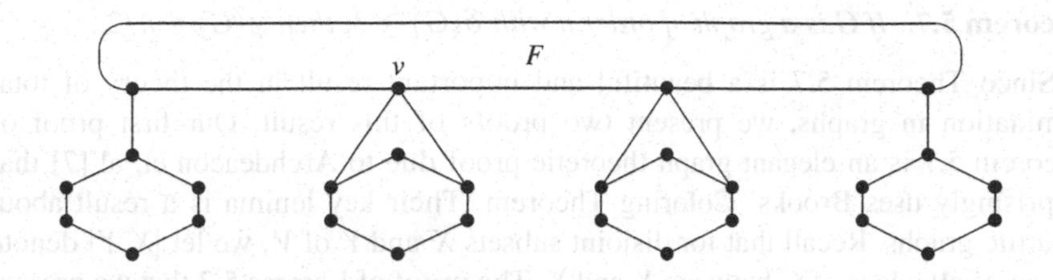

Fig. 5.2 A graph G in the family $\mathscr{F}$

v of F, add a 6-cycle and join v either to one vertex of this cycle or to two vertices at distance 2 on this cycle. Furthermore if we restrict F to be a tree, we denote the resulting family by $\mathscr{F}_T$. A graph G in the family $\mathscr{F}$ is illustrated in Fig. 5.2 (here the graph F is a 4-cycle).

In order to prove Theorem 5.5, a graph G is referred to in [102] as a $\frac{4}{7}$*-minimal graph* if G is edge minimal with respect to satisfying the following three conditions: (i) $\delta(G) \geq 2$, (ii) G is connected, and (iii) $\gamma_t(G) \geq 4n/7$, where n is the order of G. Let $\mathscr{C} = \{C_3, C_5, C_6, C_7, C_{10}, C_{14}\}$. Let H_1 be the graph obtained from a 6-cycle by adding a new vertex and joining this vertex to two vertices at distance 2 apart on the cycle. The following lemma characterizes $\frac{4}{7}$-minimal graphs.

Lemma 5.1 ([102]). *A graph G is a $\frac{4}{7}$-minimal graph if and only if $G \in \mathscr{C} \cup \mathscr{F}_T \cup \{H_1\}$.*

We note that every graph in the family $\mathscr{F}$ has total domination number exactly four-sevenths its order, as does the 7-cycle, the 14-cycle, and the graph H_1. Hence since the total domination number of a graph cannot decrease if edges are removed, it follows from Lemma 5.1 that if G satisfies the statement of Theorem 5.5, then $\gamma_t(G) \leq 4n/7$. Hence Theorem 5.5 is an immediate consequence of Lemma 5.1. The graphs of large order achieving the upper bound of Theorem 5.5 can also be characterized using Lemma 5.1.

Theorem 5.6 ([102]). *If G is a connected graph of order $n > 14$ with $\delta(G) \geq 2$ and $\gamma_t(G) = 4n/7$, then $G \in \mathscr{F}$.*

Proof. The upper bound follows immediately from Theorem 5.5. Suppose that $\gamma_t(G) = 4n/7$ and $n > 14$. Then by removing edges of G, if necessary, we produce a $\frac{4}{7}$-minimal graph G' satisfying $\gamma_t(G') = 4n/7$. By Lemma 5.1 and our earlier observations, $G' \in \mathscr{F}_T$. It can now readily be checked that $G \in \mathscr{F}$. □

5.5 Minimum Degree Three

Here we consider the case when $\delta = 3$. The upper bound in Theorem 5.5 can be improved when $\delta = 3$.

Theorem 5.7. *If G is a graph of order n with $\delta(G) \geq 3$, then $\gamma_t(G) \leq n/2$.*

Since Theorem 5.7 is a beautiful and important result in the theory of total domination in graphs, we present two proofs of this result. Our first proof of Theorem 5.7 is an elegant graph theoretic proof due to Archdeacon et. al [7] that surprisingly uses Brooks' Coloring Theorem. Their key lemma is a result about bipartite graphs. Recall that for disjoint subsets X and Y of V, we let $[X,Y]$ denote the set of all edges of G between X and Y. The proof of Lemma 5.2 that we present is from [130].

Lemma 5.2 ([7]). *Let F be a bipartite graph with partite sets (X,Y) whose vertices in Y are of degree at least 3. Then there exists a set $A \subseteq X$ of size at most $|X \cup Y|/4$ that dominates Y.*

Proof. Let $|X| = x$ and $|Y| = y$. We proceed by induction on $|V(F)| + |E(F)|$. The smallest graph described by the lemma is $K_{1,3}$, for which the statement holds. This establishes our base case. If there exists a vertex v in Y of degree at least 4, then delete any edge e incident to v. The subset A of $F - e$ guaranteed by the inductive hypothesis dominates Y in F as desired. So we may assume the vertices in Y are all of degree exactly 3.

If there exists an isolated vertex $v \in X$, then the subset A in $F - v$ guaranteed by the inductive hypothesis dominates Y in F as desired. So we may assume that each vertex in X has degree at least 1. If there exists a vertex v in X of degree at least 3, then delete that vertex and all its neighbors. Adding v into the subset A from this smaller graph yields our desired subset for F. So we may assume each vertex in X has degree at most 2.

Thus each vertex in X has degree 1 or 2 and each vertex in Y has degree 3. For $i = 1,2$, let X_i denote the vertices in X of degree i, and let $|X_i| = x_i$. Then, $x = x_1 + x_2$ and $x_1 + 2x_2 = |E(F)| = 3y$. Hence, we wish to find a set $A \subseteq X$ of size at most $(x+y)/4 = x_1/3 + 5x_2/12$ that dominates Y.

Let F' be the graph with $V(F') = X_2$ and where two vertices are adjacent in F' if and only if they have a common neighbor in F. Since each vertex in X_2 has degree 2 in F and each vertex in Y has degree 3, $\Delta(F') \leq 4$. Furthermore under our given assumptions, no component of F' is a complete graph K_5. Hence by Brooks' Coloring Theorem, F' is 4-colorable, and so F' has an independent set S of size at least $x_2/4$. Since the vertices in the set S have disjoint neighborhoods in F', $|N_F(S)| = 2|S|$. For each vertex $y \in Y \setminus N_F(S)$, we choose an adjacent vertex in $X \setminus S$ and call the resulting set of such vertices S'. Then, $A = S \cup S'$ dominates Y. Since $|S'| \leq |Y \setminus N_F(S)| = y - 2|S|$, we have $|A| = |S| + |S'| \leq y - |S| \leq y - x_2/4 = (x_1 + 2x_2)/3 - x_2/4 = x_1/3 + 5x_2/12$, as desired. □

We are now in a position to present a graph theoretic proof of Theorem 5.7. Recall its statement.

Theorem 5.7. *If G is a graph of order n with $\delta(G) \geq 3$, then $\gamma_t(G) \leq n/2$.*

Proof. Let $V(G) = \{v_1, v_2, \ldots, v_n\}$. Let F be the bipartite graph constructed from G as follows. Let $X = \{x_1, x_2, \ldots, x_n\}$ and $Y = \{y_1, y_2, \ldots, y_n\}$. For each edge $v_i v_j$ in G, add the two edges $x_i y_j$ and $x_j y_i$ in F. The bipartition of F is (X,Y). A TD-set A in G corresponds to a subset $A \subset X$ in F adjacent to every vertex in Y. The result now follows by Lemma 5.2. □

Our second proof of Theorem 5.7 that we present uses transversals in hypergraphs. We remark that the graph theoretic proof of Theorem 5.7 presented earlier has a hypergraph flavor to it (if one thinks of the hypergraph with vertex set X and edge set $\{N(y) \mid y \in Y\}$). Chvátal and McDiarmid [35] and Tuza [200] independently established the following result about transversals in hypergraphs. The proof we present is again from [130].

Theorem 5.8 ([35, 200]). *If $H = (V,E)$ is a hypergraph where all edges have size at least three, then $4\tau(H) \leq |V| + |E|$.*

Proof. It suffices for us to restrict our attention to 3-uniform hypergraphs H. We proceed by induction on the number $n = |V|$ of vertices. Let H have size $m = |E|$. If $m = 0$, then $\tau(H) = 0$ and the result follows. Hence we may assume that $m \geq 1$, and so $n \geq 3$. If $n = 3$, then $\tau(H) = 1$ and $n + m \geq 4 = 4\tau(H)$, as desired. This establishes the base case. Assume, then, that $n \geq 4$ and that all 3-uniform hypergraphs $H' = (V', E')$ where $|V'| < n$ have a transversal T' such that $4|T'| \leq |V'| + |E'|$. Let $H = (V,E)$ be a 3-uniform hypergraph where $|V| = n$ and $|E| = m$.

Suppose $d_H(v) \geq 3$ for some vertex $v \in V$. Let $H' = H - v$. Then, $H' = (V', E')$ is a 3-uniform hypergraph of order $|V'| = n - 1$ and size $|E'| \leq m - 3$. Applying the induction hypothesis to H', H' has a transversal T' such that $4|T'| \leq |V'| + |E'| \leq |V| + |E| - 4$. Thus, $T = T' \cup \{v\}$ is a transversal of H such that $4|T| \leq |V| + |E|$, as desired. Hence we may assume that $d_H(v) \leq 2$ for every vertex $v \in V$.

If some edge $\{a,b,c\} \in E$ has no vertex of degree 2, then we simply apply the induction to $H' = H - \{a,b,c\}$ to find a transversal T' such that $4|T'| \leq |V| + |E| - 4$. Thus, $T = T' \cup \{a\}$ is a transversal of H such that $4|T| \leq |V| + |E|$, as desired. Hence we may assume that every edge contains a degree-2 vertex.

Suppose some edge $\{a,b,c\} \in E$ is such that $d_H(a) = 2$ and $d_H(b) = 1$. Let $H' = H - \{a,b\}$. Then, $H' = (V', E')$ is a 3-uniform hypergraph of order $|V'| = n - 2$ and size $|E'| = m - 2$. Applying the induction hypothesis to H', H' has a transversal T' such that $4|T'| \leq |V'| + |E'| \leq |V| + |E| - 4$. Thus, $T = T' \cup \{a\}$ is a transversal of H such that $4|T| \leq |V| + |E|$, as desired. Hence we may assume that H is 2-regular; that is, $d_H(v) = 2$ for every vertex $v \in V$.

If some edge $\{a,b,c\} \in E$ appears twice in H, we simply apply the induction to $H - \{a,b,c\}$ to deduce the desired result. Hence we may assume that H has no duplicated edges.

Suppose two edges $\{a,b,c\}$ and $\{a,b,d\}$ intersect in two vertices. Let $H' = H - \{a,b\}$. Then, $H' = (V',E')$ is a 3-uniform hypergraph of order $|V'| = n-2$ and size $|E'| = m-2$. Applying the induction hypothesis to H', H' has a transversal T' such that $4|T'| \le |V'| + |E'| \le |V| + |E| - 4$. Thus, $T = T' \cup \{a\}$ is a transversal of H such that $4|T| \le |V| + |E|$, as desired.

Hence we may assume that if two edges of H intersect, then they intersect in exactly one vertex. Let $F = \{x_1,x_2,x_3\}$ be any edge in H. For $i = 1,2,3$, let F_i be the edge in H, distinct from F, such that $x_i \in F_i$. From our earlier assumptions, we note that for $1 \le i < j \le 3$, $|(F_i \cup F_j) \setminus \{x_1,x_2,x_3\}| \ge 3$. Let F_{ij} be any 3-element subset of $(F_i \cup F_j) \setminus \{x_1,x_2,x_3\}$. Let H' be obtained from the hypergraph $H - \{x_1,x_2,x_3\}$ by adding the edges F_{12}, F_{13} and F_{23}. Then, $H' = (V',E')$ is a 3-uniform hypergraph of order $|V'| = n-3$ and size $|E'| = m-1$. Applying the induction hypothesis to H', H' has a transversal T' such that $4|T'| \le |V'| + |E'| \le |V| + |E| - 4$.

If T' contains at least one vertex from each of the deleted edges F_1, F_2, and F_3 (i.e., if T' covers each of F_1, F_2, and F_3), then we need only add to T' any vertex from the edge F to form a transversal T of H satisfying $4|T| = 4|T'| + 4 \le |V| + |E|$. Hence we may assume that at least one of F_1, F_2, and F_3, say, F_1, is not covered by T' in H; that is, we may assume $T' \cap F_1 = \emptyset$. But then $T' \cap F_{12} \subseteq F_2$ and $T' \cap F_{13} \subseteq F_3$. Thus, F_2 and F_3 are covered by T' in H. Therefore, $T = T' \cup \{x_1\}$ is a transversal in H such that $4|T| \le |V| + |E|$, as desired. □

We are now in a position to present a hypergraph proof of Theorem 5.7. Recall its statement.

Theorem 5.7. *If G is a graph of order n with $\delta(G) \ge 3$, then $\gamma_t(G) \le n/2$.*

Proof. Let G be a graph of order n with $\delta(G) \ge 3$, and let H_G be the *ONH* of G. Then, each edge of H_G has size at least 3. By Theorem 5.8, there exists a transversal in H_G of size at most $(n+n)/4 = n/2$. Hence, $\gamma_t(G) = \tau(H_G) \le n/2$. □

Can we characterize the graphs achieving the upper bound of Theorem 5.7? The generalized Petersen graph G_{16} of order 16 shown in Fig. 1.2 achieves equality in Theorem 5.7. Two infinite families $\mathcal{G}$ and $\mathcal{H}$ of connected cubic graphs (described below) with total domination number one-half their orders are constructed in [66] which shows that the bound of Theorem 5.7 is sharp. For $k \ge 1$, let G_k be the graph constructed as follows. Consider two copies of the path P_{2k} with respective vertex sequences $a_1b_1a_2b_2\ldots a_kb_k$ and $c_1d_1c_2d_2\ldots c_kd_k$. Let $A = \{a_1,a_2,\ldots,a_k\}$, $B = \{b_1,b_2,\ldots,b_k\}$, $C = \{c_1,c_2,\ldots,c_k\}$, and $D = \{d_1,d_2,\ldots,d_k\}$. For each $i \in \{1,2,\ldots,k\}$, join a_i to d_i and b_i to c_i. To complete the construction of the graph G_k, join a_1 to c_1 and b_k to d_k. Let $\mathcal{G} = \{G_k \mid k \ge 1\}$. For $k \ge 2$, let H_k be obtained from G_k by deleting the two edges a_1c_1 and b_kd_k and adding the two edges a_1b_k and c_1d_k. Let $\mathcal{H} = \{H_k \mid k \ge 2\}$. We note that G_k and H_k are cubic graphs of order $4k$. Further, we note that $G_1 = K_4$. The graphs $G_4 \in \mathcal{G}$ and $H_4 \in \mathcal{H}$, for example, are illustrated in Fig. 5.3.

Hence we have two infinite families of connected graphs that achieve the upper bound of Theorem 5.7, as well as one connected graph of order 16. Surprisingly,

Fig. 5.3 Cubic graphs $G_4 \in \mathscr{G}$ and $H_4 \in \mathscr{H}$

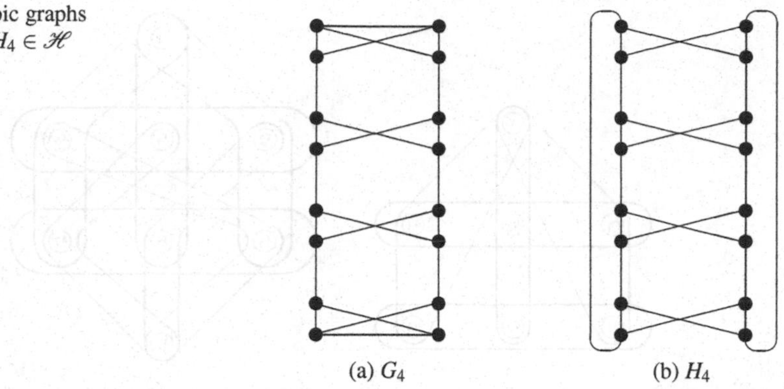

there are no other extremal connected graphs as shown in [133]. In an attempt to characterize the extremal graphs of Theorem 5.7, the authors in [133] first characterized the hypergraphs that achieve the upper bound of Theorem 5.8. For this purpose, we will define a number of classes of hypergraphs, as well as a couple of finite hypergraphs (Figs. 5.4–5.8).

Definition 5.1. Let H_7 be the hypergraph with vertex set and edge set given by

$$\begin{aligned} V(H_7) &= \{x, a_1, b_1, a_2, b_2, a_3, b_3\} \\ E(H_7) &= \{\{x, a_1, b_1\}, \{x, a_2, b_2\}, \{x, a_3, b_3\}, \{a_1, a_2, a_3\}, \{b_1, b_2, b_3\}\}. \end{aligned}$$

Definition 5.2. Let H_8 be the hypergraph with vertex set $V(H_8) = \{x_1, x_2, \ldots, x_8\}$ and the following edge set:

$$\begin{aligned} E(H_8) = \{&\{x_1,x_2,x_3\},\{x_1,x_4,x_5\},\{x_1,x_6,x_7\},\{x_2,x_4,x_6\}, \\ &\{x_3,x_5,x_7\},\{x_2,x_5,x_8\},\{x_3,x_6,x_8\},\{x_4,x_7,x_8\}\}. \end{aligned}$$

Definition 5.3. Let $i \geq 0$ be an arbitrary integer. Let $H_i^{d=1}$ be the 3-uniform hypergraph defined as follows. Let $V(H_i^{d=1}) = \{u, x_0, x_1, \ldots, x_i, y_0, y_1, \ldots, y_i\}$. Let the edge set of $H_i^{d=1}$ be defined as follows:

$$E(H_i^{d=1}) = \{\{u, x_0, y_0\}\} \cup \left(\bigcup_{a=1}^{i} \{\{x_{a-1}, x_a, y_a\}, \{y_{a-1}, x_a, y_a\}\}\right).$$

Let $H^{d=1} = \{H_0^{d=1}, H_1^{d=1}, H_2^{d=1}, \ldots\}$.

Definition 5.4. Let $i, j \geq 1$ be arbitrary integers. Let $H_{i,j}^*$ be the hypergraph defined as follows;

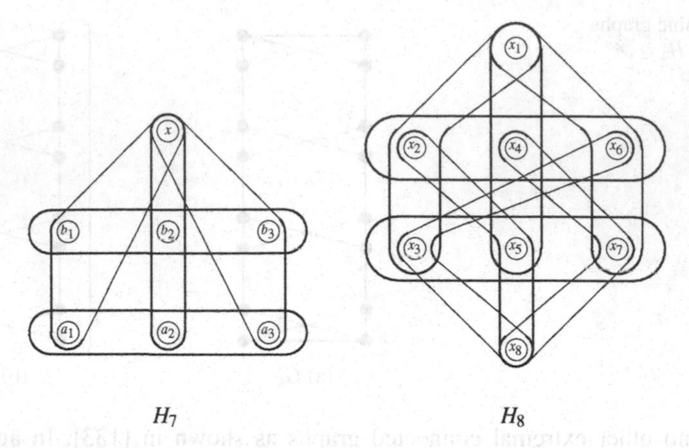

Fig. 5.4 The hypergraphs H_7 and H_8

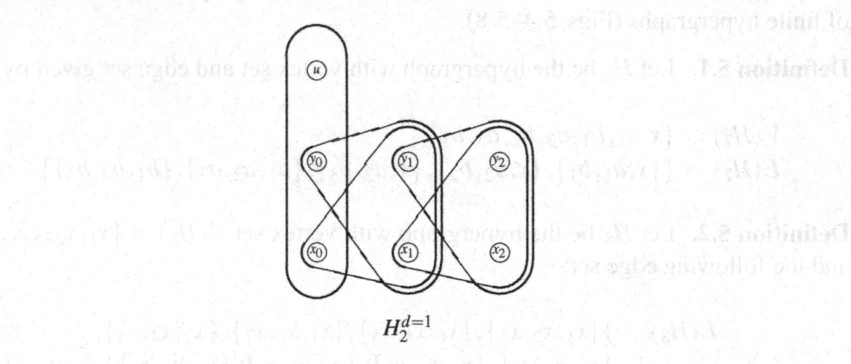

Fig. 5.5 The hypergraph $H_2^{d=1}$

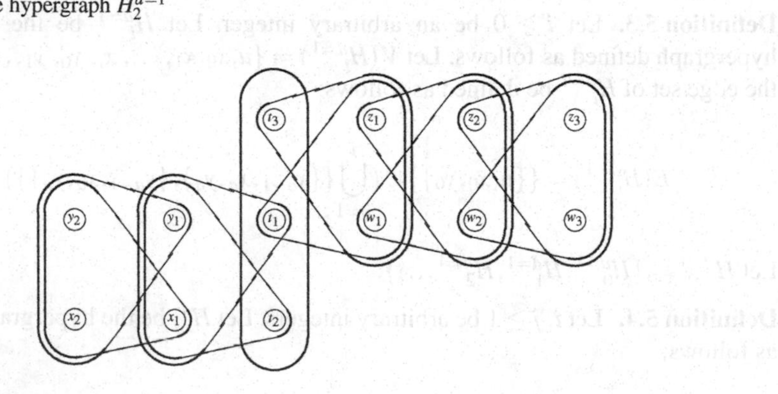

Fig. 5.6 The hypergraph $H^*_{2,3}$

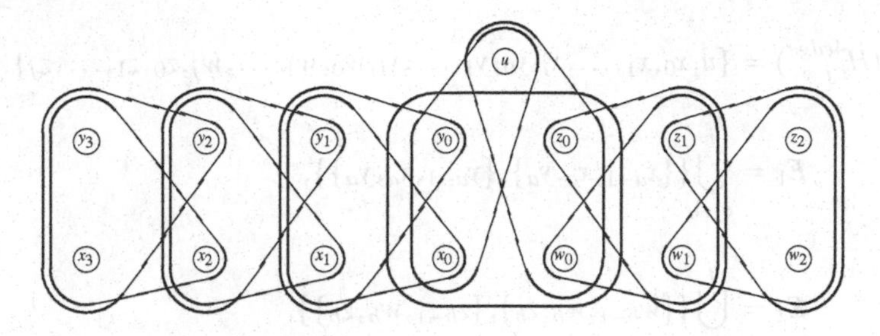

Fig. 5.7 The hypergraph $H_{3,2}^{4edge}$

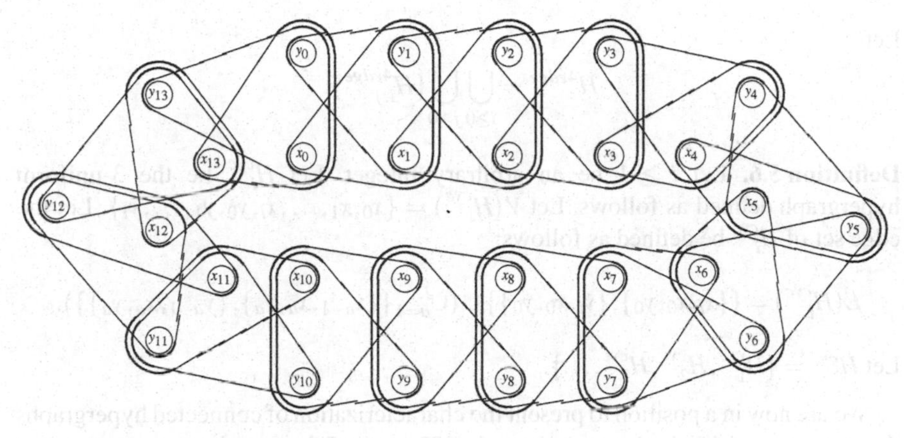

Fig. 5.8 The hypergraph H_{13}^{cyc}

$$V(H_{i,j}^{*}) = \{t_1,t_2,t_3,x_1,x_2,\ldots,x_i,y_1,y_2,\ldots,y_i,w_1,w_2,\ldots,w_j,z_1,z_2,\ldots,z_j\},$$

$$E_1 = \{\{t_1,x_1,y_1\},\{t_2,x_1,y_1\}\} \cup (\bigcup_{a=2}^{i}\{\{x_{a-1},x_a,y_a\},\{y_{a-1},x_a,y_a\}\}),$$

$$E_2 = \{\{t_1,w_1,z_1\},\{t_3,w_1,z_1\}\} \cup (\bigcup_{b=2}^{j}\{\{w_{b-1},w_b,z_b\},\{z_{b-1},w_b,z_b\}\}),$$

$$E(H_{i,j}^{*}) = \{\{t_1,t_2,t_3\}\} \cup E_1 \cup E_2.$$

Let

$$H^{*} = \bigcup_{i\geq 1}\bigcup_{j\geq 1}\{H_{i,j}^{*}\}.$$

Definition 5.5. Let $i,j \geq 0$ be arbitrary integers. Let $H_{i,j}^{4edge}$ be the hypergraph defined as follows:

$$V(H_{i,j}^{4edge}) = \{u, x_0, x_1, \ldots, x_i, y_0, y_1, \ldots, y_i, w_0, w_1, \ldots, w_j, z_0, z_1, \ldots, z_j\},$$

$$E_1 = \bigcup_{a=1}^{i} \{\{x_{a-1}, x_a, y_a\}, \{y_{a-1}, x_a, y_a\}\},$$

$$E_2 = \bigcup_{b=1}^{j} \{\{w_{b-1}, w_b, z_b\}, \{z_{b-1}, w_b, z_b\}\},$$

$$E(H_{i,j}^{4edge}) = \{\{u, x_0, y_0\}, \{u, w_0, z_0\}, \{x_0, y_0, z_0, w_0\}\} \cup E_1 \cup E_2.$$

Let

$$H^{4edge} = \bigcup_{i \geq 0} \bigcup_{j \geq 0} \{H_{i,j}^{4edge}\}.$$

Definition 5.6. Let $i \geq 1$ be an arbitrary integer. Let H_i^{cyc} be the 3-uniform hypergraph defined as follows. Let $V(H_i^{cyc}) = \{x_0, x_1, \ldots, x_i, y_0, y_1, \ldots, y_i\}$. Let the edge set of H_i^{cyc} be defined as follows:

$$E(H_i^{cyc}) = \{\{x_i, x_0, y_0\}, \{y_i, x_0, y_0\}\} \cup (\cup_{a=1}^{i} \{\{x_{a-1}, x_a, y_a\}, \{y_{a-1}, x_a, y_a\}\}).$$

Let $H^{cyc} = \{H_1^{cyc}, H_2^{cyc}, H_3^{cyc}, \ldots\}$.

We are now in a position to present the characterization of connected hypergraphs that achieve equality in the upper bound of Theorem 5.8.

Theorem 5.9 ([133]). *Let H be a connected hypergraph where all edges have size at least three with n vertices and m edges. If $4\tau(H) = n + m$, then $H \in \{H_7, H_8\} \cup H^{cyc} \cup H^{d=1} \cup H^* \cup H^{4edge}$.*

The details of a proof of this theorem can be found in [133]. As a consequence of this characterization, it is now possible to characterize the connected graphs with minimum degree at least three that achieve equality in the bound of Theorem 5.7.

Theorem 5.10 ([133]). *If G is a connected graph with minimum degree at least three and total domination number one-half its order, then $G \in \mathcal{G} \cup \mathcal{H}$ or G is the generalized Petersen graph of order 16 shown in Fig. 1.2.*

We next present an outline of the proof of Theorem 5.10 that can be found in [130]. Let G be a connected graph with $\gamma_t(G) = n(G)/2$, and let H_G be the *ONH* of G. Using Theorem 5.9, we first show that G is 3-regular and that every component in H_G is equal to H_8 or belongs to H^{cyc}. Applying the result of Theorem 1.1, we show that if G is not bipartite, then G must contain a 3-cycle and we deduce that $H_G \in H^{cyc}$ and $G \in \mathcal{G}$. Further we show that if G is bipartite, then H_G contains two components. If a component of H_G is equal to H_8, then we show that G is the generalized Petersen graph of order 16 shown in Fig. 1.2, while if a component of

H_G belongs to H^{cyc}, then we show that $G \in \mathscr{H}$. This completes the outline of a proof of Theorem 5.10.

The result of Theorem 5.7 was strengthened by Lam and Wei [157].

Theorem 5.11 ([157]). *If G is a graph of order n with $\delta(G) \geq 2$ such that every component of the subgraph of G induced by its set of degree-2 vertices has order at most two, then $\gamma_t(G) \leq n/2$.*

As a special case of their result, we have the following:

Theorem 5.12 ([157]). *If G is a graph of order n with $\delta(G) \geq 2$ such that $d(u) + d(v) \geq 5$ for every two adjacent vertices u and v of G, then $\gamma_t(G) \leq n/2$.*

The result of Theorem 5.11 is generalized further in [131]. Let G be a connected graph of order n with minimum degree at least two that is not 2-regular. We define a vertex as large if it has degree more than 2, and we let $\mathscr{L}$ be the set of all large vertices of G. Let P be any component of $G - \mathscr{L}$; it is a path. If $|V(P)| \equiv 0 \pmod 4$ and either the two ends of P are adjacent in G to the same large vertex or the two ends of P are adjacent to different, but adjacent, large vertices in G, we call P a 0^*-path. If $|V(P)| \geq 5$ and $|V(P)| \equiv 1 \pmod 4$ with the two ends of P adjacent in G to the same large vertex, we call P a 1^*-path. If $|V(P)| \equiv 3 \pmod 4$, we call P a 3^*-path. For $i \in \{0,1,3\}$, we denote the number of i^*-paths in G by p_i.

Theorem 5.13 ([131]). *If G is a connected graph of order n with $\delta(G) \geq 2$ and $\Delta(G) \geq 3$, then $\gamma_t(G) \leq (n + p_0 + p_1 + p_3)/2$.*

Suil O and Douglas West [192] define a *balloon* in a graph G to be a maximal 2-edge-connected subgraph incident to exactly one cut edge of G, and they let $b(G)$ be the number of balloons in G. The following result improves the upper bound of Theorem 5.7 when G is a cubic graph having at least one balloon.

Theorem 5.14 ([192]). *If G is a connected cubic graph of order n, then $\gamma_t(G) \leq n/2 - b(G)/2$, except that $\gamma_t(G) \leq n/2 - 1$ when $b(G) = 3$ and the three balloons have a common neighbor, and this is sharp for all even values of $b(G)$.*

5.6 Minimum Degree Four

In this section, we consider the case when $\delta = 4$. It is a natural question to ask if the upper bound in Theorem 5.7 can be improved if we restrict the minimum degree to be four. This is indeed the case. However in order to improve the bound of Theorem 5.7, we make a transition from total domination in graphs to transversals in hypergraphs since it appears difficult to improve the bound using graph theoretic techniques.

We shall adopt the following notation. Let H_1 be a hypergraph with vertices $\{y_1, y_2\}$ and one edge $\{y_1, y_2\}$. Let H_2 be a hypergraph with vertices $\{x_1, x_2, x_3, x_4, x_5\}$ and edges $\{\{x_1, x_2, x_3\}, \{x_1, x_4, x_5\}, \{x_2, x_3, x_4, x_5\}\}$.

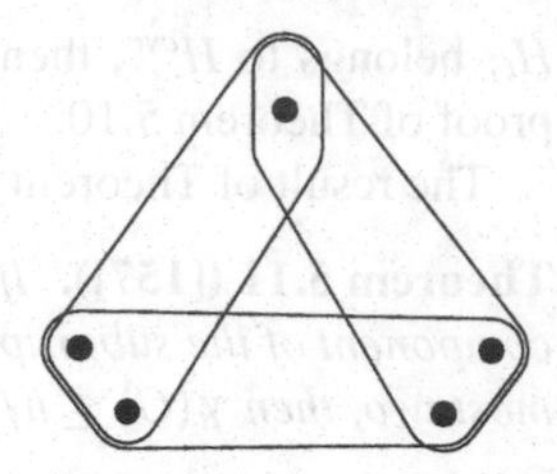

Fig. 5.9 The hypergraph H_2

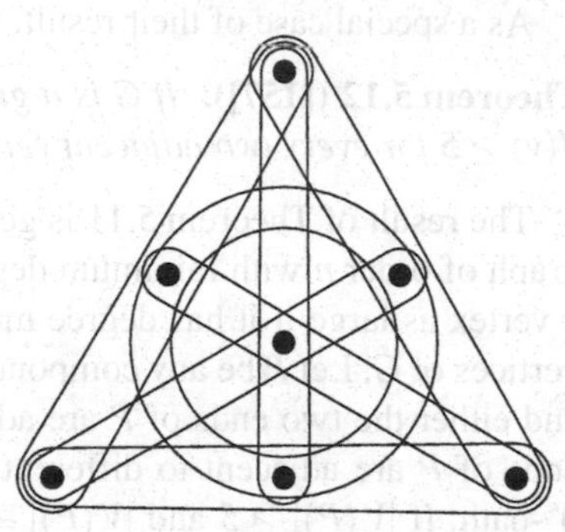

Fig. 5.10 The Fano plane

The hypergraph H_2 is illustrated in Fig. 5.9, albeit without the vertex labels. For an arbitrary hypergraph H, we let $h(H)$ denote the number of H_1-components and H_2-components in H. Further for $i = 2,3,4$, let $e_i(H)$ denote the number of edges in H of size i.

Theorem 5.15 ([197,208]). *If H is a hypergraph with n vertices and with edges of size at most four and at least two, then there exists a transversal T of H such that*

$$21|T| \le 5|V(H)| + 10e_2(H) + 6e_3(H) + 4e_4(H) + h(H).$$

As a special case of Theorem 5.15, we have the following result due to Thomasse and Yeo [197].

Theorem 5.16 ([197]). *Every 4-uniform hypergraph on n vertices and m edges has a transversal with no more than $(5n+4m)/21$ vertices.*

A characterization of the 4-uniform hypergraphs that achieve equality in the bound of Theorem 5.16 has recently been given in [208]. Recall that the Fano plane is the hypergraph shown in Fig. 5.10.

Theorem 5.17 ([208]). *If H is a connected 4-uniform hypergraph, different from the complement of the Fano plane, on n vertices and m edges, then*

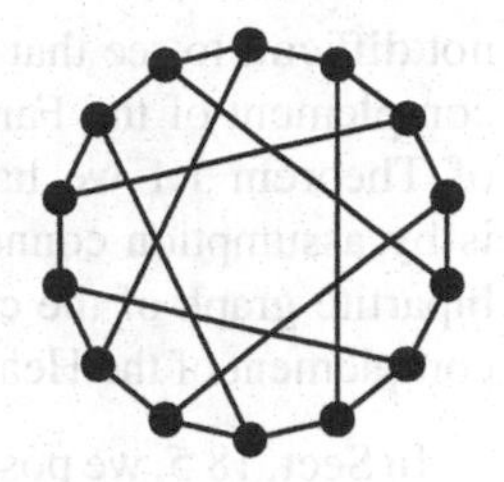

Fig. 5.11 The Heawood graph

$$\tau(H) \le \frac{1}{21}\left[\left(5+\frac{39}{283}\right)n+\left(4-\frac{40}{283}\right)m\right].$$

As a consequence of Theorems 5.16 and 5.17, we have the following result on the total domination number of a graph with minimum degree at least four. Recall that the Heawood graph is the graph shown in Fig. 5.11 (which is the unique 6-cage). The bipartite complement of the Heawood graph can also be seen as the incidence bipartite graph of the complement of the Fano plane.

Theorem 5.18 ([197, 208]). *If G is a connected graph of order n with $\delta(G) \ge 4$, then $\gamma_t(G) \le 3n/7$, with equality if and only if G is the bipartite complement of the Heawood graph.*

Proof. Let H_G be the *ONH* of G. Then, each edge of H_G has size at least 4. Let H be obtained from H_G by shrinking all edges of H_G, if necessary, to edges of size 4. Then, H is a 4-uniform hypergraph with n vertices and n edges. By Theorem 5.16, $\tau(H) \le (5n+4n)/21 = 3n/7$. Hence, $\gamma_t(G) = \tau(H_G) \le \tau(H) \le 3n/7$. This establishes the desired upper bound.

Suppose next that $\gamma_t(G) = 3n/7$. Let $H_1, H_2, \dots, H_k$ be the components of H. For $i = 1, 2, \dots, k$, let H_i have order n_i and size m_i. Then, $\sum_{i=1}^{k} n_i = n$ and $\sum_{i=1}^{k} m_i = n$. By Theorem 5.16, we have $\tau(H_i) \le (5n_i + 4m_i)/21$ for all $i = 1, 2, \dots, k$. By Theorem 1.2, we therefore have that

$$\frac{3n}{7} = \gamma_t(G) = \tau(H_G) \le \tau(H) = \sum_{i=1}^{k} \tau(H_i) \le \sum_{i=1}^{k} \left(\frac{5n_i+4m_i}{21}\right) = \frac{3n}{7}. \tag{5.1}$$

Hence we have equality throughout the inequality chain (5.1), implying that $\tau(H_G) = \tau(H)$ and $\tau(H_i) = (5n_i + 4m_i)/21$ for all $i = 1, 2, \dots, k$. Recall that since H has the same size as the *ONH*, H_G, of G, we have that $m(H) = \sum_{i=1}^{k} m_i = n$. If $m_i \ge n_i$ for some $i \in \{1, 2, \dots, k\}$, then by Theorem 5.17, H_i is the complement of the Fano plane, for otherwise $\tau(H_i) < (5n_i + 4m_i)/21$, a contradiction. Further since the complement of the Fano plane has an equal number of vertices and edges, this implies that $m_i = n_i$. In particular, $m_i > n_i$ is not possible. This in turn implies that if $m_i < n_i$ for some $i \in \{1, 2, \dots, k\}$, then $n = \sum_{i=1}^{k} m_i < \sum_{i=1}^{k} n_i = n$, which is not possible. Therefore, $m_i = n_i$ for all $i \in \{1, 2, \dots, k\}$. This implies that every component H_i of H is the complement of the Fano plane. If $H \ne H_G$, then it is not difficult to see that $\tau(H) < \tau(H_G)$, a contradiction. Hence, $H = H_G$. Since the complement of the Fano plane is not the *ONH* of any graph, applying the result of Theorem 1.1 we have that H consists of precisely two components since G is by assumption connected. It is now not difficult to see that G is the incidence bipartite graph of the complement of the Fano plane or, equivalently, the bipartite complement of the Heawood graph. □

In Sect. 18.5, we pose the following conjecture: If G is connected graph of order n with $\delta(G) \geq 4$ that is not the bipartite complement of the Heawood graph, then $\gamma_t(G) \leq 2n/5$. A discussion on the sharpness of this conjecture, if true, can be found in Sect. 18.5.

5.7 Minimum Degree Five

In this section, we consider the case when $\delta = 5$. Can the upper bound in Theorem 5.18 be improved if we restrict the minimum degree to be five? Let H_1 and H_2 be the hypergraphs defined in Sect. 5.6. For $i \in \{1,2\}$, let $h_i(H)$ be the number of components of H, which are isomorphic to H_i. Let $n(H)$ and $m(H)$ denote the number of vertices and edges in H, respectively. For $i \in \{2,3,4,5\}$, let $e_i(H)$ denote the number of edges of H of size i. The following result is established in [57].

Theorem 5.19 ([57]). *Let H be a hypergraph with n vertices and with edges of size at least two and at most five. Then there exists a transversal T such that*

$$132|T| \leq 30n(H) + 69e_2(H) + 42e_3(H) + 28e_4(H) + 21e_5(H) + 3h_1(H) + 2h_2(H).$$

As an immediate consequence of Theorem 5.19, we have the following corollaries.

Theorem 5.20 ([57]). *If H is a 5-uniform hypergraph on n vertices and m edges, then $\tau(H) \leq (10n + 7m)/44$.*

Theorem 5.21 ([57]). *If G is a connected graph of order n with $\delta(G) \geq 5$, then $\gamma_t(G) \leq 17n/44$.*

It is unlikely that the bounds of Theorems 5.20 and 5.21 are best possible. In Sect. 18.6, we pose the following conjecture: If H is a 5-uniform hypergraph on n vertices and m edges, then $\tau(H) \leq (9n + 7m)/44$. This conjecture, if true, would prove two conjectures due to Thomassé and Yeo [197]. Their first conjecture states that if H is a 5-uniform hypergraph on n vertices and n edges, then $\tau(H) \leq 4n/11$. Their second related conjecture states that a connected graph G of order n with $\delta(G) \geq 5$ satisfies $\gamma_t(G) \leq 4n/11$. For a more detailed discussion of these conjectures, we refer the reader to Sect. 18.6.

Table 5.1 Upper bounds on the total domination number of a graph G

$\delta(G) \geq 1 \Rightarrow \gamma_t(G) \leq \frac{2}{3}n$	if $n \geq 3$ and G is connected
$\delta(G) \geq 2 \Rightarrow \gamma_t(G) \leq \frac{4}{7}n$	if $n \geq 11$ and G is connected
$\delta(G) \geq 3 \Rightarrow \gamma_t(G) \leq \frac{1}{2}n$	
$\delta(G) \geq 4 \Rightarrow \gamma_t(G) \leq \frac{3}{7}n$	
$\delta(G) \geq 5 \Rightarrow \gamma_t(G) \leq \frac{17}{44}n$	
Any $\delta(G) \geq 1 \Rightarrow \gamma_t(G) \leq \left(\frac{1+\ln\delta}{\delta}\right) n$	

5.8 Summary of Known Results on Bounds in Terms of Order

We summarize the known upper bounds on the total domination number of a graph G in terms of its order n and minimum degree δ in Table 5.1.

5.9 Total Domination and Connectivity

In this section we investigate upper bounds on the total domination number of a graph with high connectivity. In particular, we consider the following two problems.

Problem 5.3. For a k-connected graph G of large order n, where $k \geq 1$, find a sharp upper bound $g(k,n)$ on $\gamma_t(G)$ in terms of k and n.

Problem 5.4. Characterize the k-connected graphs G of large order n satisfying $\gamma_t(G) = g(k,n)$.

If $k = 1$, then G is a connected graph and the desired results are given by Theorems 5.2 and 5.3. Hence in what follows we restrict our attention to $k \geq 2$.

In Sect. 5.4 we established bounds on the total domination number of a connected graph with minimum degree two. Every graph in the family $\mathscr{F}$ of extremal graphs of large order that achieve equality in the bound of Theorem 5.6 has a cut vertex. It is therefore a natural question to ask whether this upper bound of $4n/7$ can be improved if we restrict our attention to 2-connected graphs. Indeed if G has sufficiently large order, then the bound can be improved.

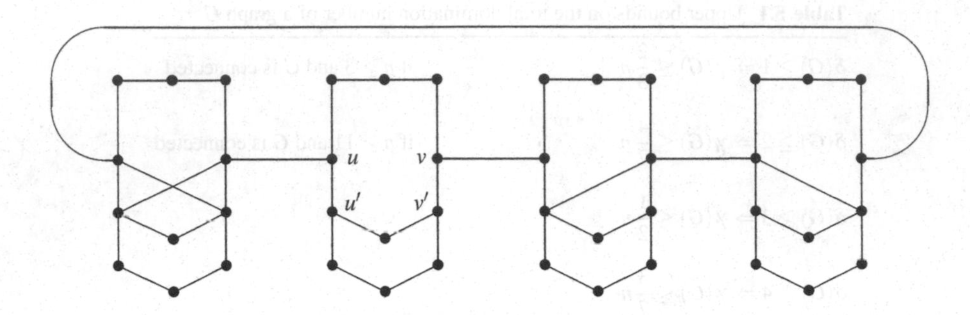

Fig. 5.12 A graph in the family $\mathscr{I}$

Theorem 5.22 ([134]). *If G is a 2-connected graph of order $n \geq 19$, then $\gamma_t(G) \leq 6n/11$.*

To illustrate the sharpness of Theorem 5.22, let $r \geq 2$ be an integer and let $\mathscr{I}$ be the family of all graphs that can be obtained from a 2-connected graph H of order $2r$ that contains a perfect matching M as follows. For each edge $e = uv$ in the matching M, subdivide the edge e three times, add a 6-cycle, select two vertices u' and v' at distance 2 apart on this cycle, and join u to u' and v to v'. The edges uv' and $u'v$ are optional edges that may be added. Let G denote the resulting graph of order $n = 11r$. Then, $\gamma_t(G) = 6r = 6n/11$. A graph in the family $\mathscr{I}$ with $r = 4$ that is obtained from an 8-cycle H is shown in Fig. 5.12.

Next we consider graphs with large connectivity. In Sect. 5.5, we established bounds on the total domination number of a connected graph with minimum degree three. As a consequence of Theorems 5.7 and 5.10, we have the following result. Recall that the family $\mathscr{H}$ of cubic graphs is defined in Sect. 5.5.

Theorem 5.23 ([133]). *If G is a 3-connected graph of order n, then $\gamma_t(G) \leq n/2$ with equality if and only if $G = K_4$ or $G \in \mathscr{H}$ or G is the generalized Petersen graph of order 16 shown in Fig. 1.2.*

For 4-connected graphs, we have the following result which is an immediate consequence of Theorem 5.18 in Sect. 5.6 which establishes an upper bound on the total domination number of a connected graph with minimum degree four. Recall that a drawing of the Heawood graph is given in Fig. 5.11.

Theorem 5.24 ([208]). *If G is a 4-connected graph of order n, then $\gamma_t(G) \leq 3n/7$, with equality if and only if G is the relative complement of the Heawood graph.*

We remark that Problem 5.3 and Problem 5.4 have yet to be solved for any value of $k \geq 5$. The best upper bounds to date for a k-connected graph are the upper bounds established in Chap. 5 for graphs with minimum degree at least k.

Chapter 6
Total Domination in Planar Graphs

6.1 Introduction

The decision problem to determine the total domination number of a graph remains NP-hard even when restricted to planar graphs of maximum degree three [71]. Hence it is of interest to determine upper bounds on the total domination number of a planar graph.

For $k \geq 1$ an integer, if T is a tree obtained from a star $K_{1,k}$ by subdividing every edge once, then $\text{rad}(T) = 2$ and $\text{diam}(T) = 4$ and $\gamma_t(T) = k+1$. Hence a tree of radius 2 and diameter 4 can have arbitrarily large total domination number. So the interesting question is what happens when the diameter is 2 or 3. As pointed out in [164], the restriction of bounding the diameter of a planar graph is reasonable to impose because planar graphs with small diameter are often important in applications (see [68]).

6.2 Diameter Two Planar Graphs

MacGillivray and Seyffarth [164] proved that planar graphs with diameter two or three have bounded domination numbers. In particular, this implies that the domination number of such a graph can be determined in polynomial time.

Theorem 6.1 ([164]). *If G is a planar graph with* $\text{diam}(G) = 2$, *then* $\gamma(G) \leq 3$.

The bound of Theorem 6.1 is sharp as may be seen by considering the graph G_9 of Fig. 6.1 constructed by MacGillivray and Seyffarth [164]. The graph G_9 of Fig. 6.1 is in fact the unique planar graph of diameter two with domination number 3 as shown in [75].

Theorem 6.2 ([75]). *If G is a planar graph with* $\text{diam}(G) = 2$, *then* $\gamma(G) \leq 2$ *or $G = G_9$ where G_9 is the graph of Fig. 6.1.*

M.A. Henning and A. Yeo, *Total Domination in Graphs*, Springer Monographs in Mathematics, DOI 10.1007/978-1-4614-6525-6_6,

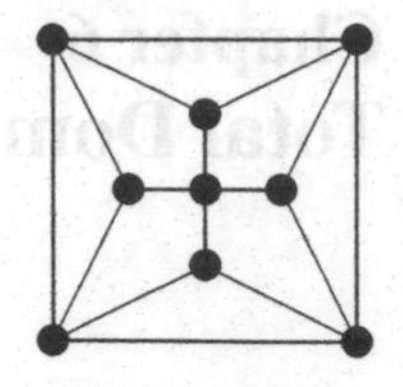

Fig. 6.1 A planar graph G_9 of diameter 2 with domination number 3

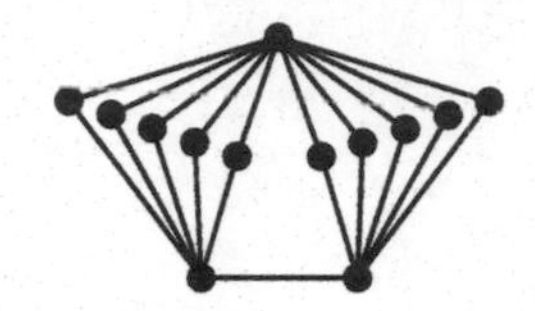

Fig. 6.2 A graph in the family $\mathscr{G}_{\text{planar}}$

Let G be a planar graph with $\text{diam}(G) = 2$. If $G = G_9$ where G_9 is the graph of Fig. 6.1, then $\gamma_t(G) = 3$. If $G \neq G_9$, then by Theorem 6.2, $\gamma(G) \leq 2$. Let $\{u, v\}$ be a dominating set in G. If u and v are adjacent, then $\{u, v\}$ is a TD-set in G, and so $\gamma_t(G) \leq 2$. If u and v are not adjacent, then since G has diameter-2, there is a common neighbor, w, say, of u and v. Thus, $\{u, v, w\}$ is a TD-set in G, and so $\gamma_t(G) \leq 3$. Hence as an immediate consequence of Theorem 6.2, we have the following result.

Theorem 6.3 ([75]). *If G is a planar graph with* $\text{diam}(G) = 2$, *then* $\gamma_t(G) \leq 3$.

A characterization of planar graphs with diameter two and total domination number three seems difficult to obtain since there are infinitely many such graphs. For an example of one such infinite family, consider the family $\mathscr{G}_{\text{planar}}$ of planar graphs with diameter two and total domination number three that can be obtained from a 5-cycle $v_1v_2v_3v_4v_5v_1$ by replacing the vertices v_2 and v_5 with nonempty independent set A_2 and A_5, respectively, and adding all edges between A_2 and $\{v_1, v_3\}$ and adding all edges between A_5 and $\{v_1, v_4\}$. A graph in the family $\mathscr{G}_{\text{planar}}$ with $|A_2| = |A_5| = 5$ is shown in Fig. 6.2 (where the labels of the vertices are omitted).

If we restrict our attention to planar graphs with certain structural properties, then a characterization of such planar graphs with diameter two and total domination number three is possible. Along these lines, the authors in [121] say that a graph G satisfies the *domination-cycle property* if there is some $\gamma(G)$-set not contained in any induced 5-cycle of G.

Let F_1 be the graph shown in Fig. 6.3 with three specified vertices c_1, c_2, and c_3 as indicated. Let $E_1^* = \{c_1c_2, c_1c_3, c_2c_3\}$. Let $\mathscr{F}_1$ be the family of eight graphs defined by $\mathscr{F}_1 = \{F \mid V(F) = V(F_1) \text{ and } E(F) = E(F_1) \cup E_1 \text{ where } E_1 \text{ is any subset of } E_1^*\}$.

Let F_2 be the graph shown in Fig. 6.3 with four specified vertices c_1, c_2, c_3, and c_4 as indicated. Let $E_2^* = \{c_1c_2, c_2c_3, c_3c_4, c_4c_1\}$. Let $\mathscr{F}_2$ be the family of sixteen graphs defined by $\mathscr{F}_2 = \{F \mid V(F) = V(F_2) \text{ and } E(F) = E(F_2) \cup E_2 \text{ where } E_2 \text{ is any subset of } E_2^*\}$.

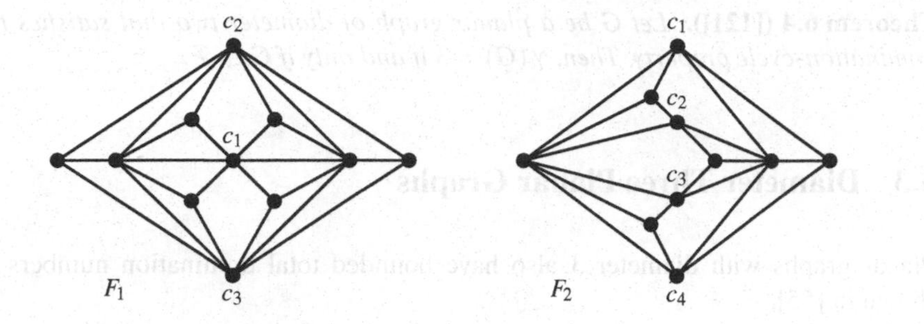

Fig. 6.3 The graphs F_1 and F_2

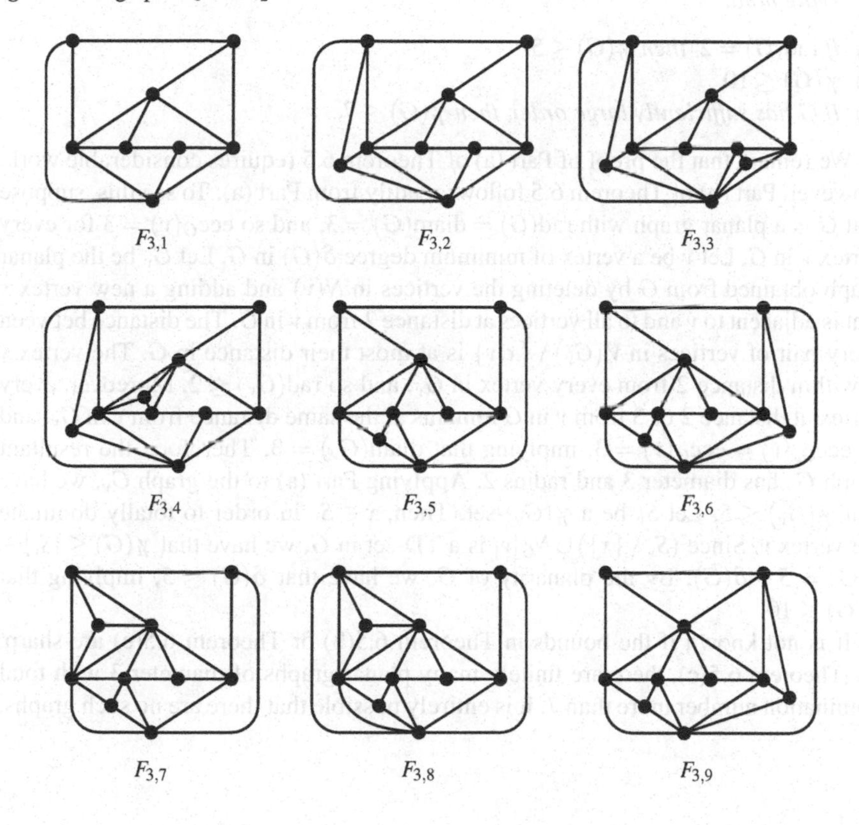

Fig. 6.4 The family $\mathscr{F}_3$

Let $\mathscr{F}_3$ be the family of nine graphs shown in Fig. 6.4. Let $\mathscr{F}$ be the family of thirty-four graphs defined by $\mathscr{F} = \mathscr{F}_1 \cup \mathscr{F}_2 \cup \mathscr{F}_3 \cup \{G_9\}$, where G_9 is the graph of Fig. 6.1.

The planar graphs with diameter two and total domination number three that satisfy the domination-cycle property are characterized in [121].

Theorem 6.4 ([121]). *Let G be a planar graph of diameter two that satisfies the domination-cycle property. Then, $\gamma_t(G) = 3$ if and only if $G \in \mathscr{F}$.*

6.3 Diameter Three Planar Graphs

Planar graphs with diameter 3 also have bounded total domination numbers as shown in [55].

Theorem 6.5 ([55]). *Let G be a planar graph of diameter at most 3. Then the following hold:*

(a) *If* $\text{rad}(G) = 2$, *then* $\gamma_t(G) \le 5$.
(b) $\gamma_t(G) \le 10$.
(c) *If G has sufficiently large order, then* $\gamma_t(G) \le 7$.

We remark that the proof of Part (a) of Theorem 6.5 requires considerable work. However, Part (b) of Theorem 6.5 follows readily from Part (a). To see this, suppose that G is a planar graph with $\text{rad}(G) = \text{diam}(G) = 3$, and so $\text{ecc}_G(v) = 3$ for every vertex v in G. Let v be a vertex of minimum degree $\delta(G)$ in G. Let G_v be the planar graph obtained from G by deleting the vertices in $N(v)$ and adding a new vertex x that is adjacent to v and to all vertices at distance 2 from v in G. The distance between every pair of vertices in $V(G_v) \setminus \{x,v\}$ is at most their distance in G. The vertex x is within distance 2 from every vertex in G_v, and so $\text{rad}(G_v) \le 2$. Moreover, every vertex at distance 2 or 3 from v in G remains at the same distance from v in G_v, and so $\text{ecc}_{G_v}(v) = \text{ecc}_G(v) = 3$, implying that $\text{diam}(G_v) = 3$. Therefore, the resultant graph G_v has diameter 3 and radius 2. Applying Part (a) to the graph G_v, we have that $\gamma_t(G_v) \le 5$. Let S_v be a $\gamma_t(G_v)$-set. Then, $x \in S_v$ in order to totally dominate the vertex v. Since $(S_v \setminus \{x\}) \cup N_G[v]$ is a TD-set in G, we have that $\gamma_t(G) \le |S_v| + \delta(G) = 5 + \delta(G)$. By the planarity of G, we have that $\delta(G) \le 5$, implying that $\gamma_t(G) \le 10$.

It is not known if the bounds in Theorem 6.5(b) or Theorem 6.5(c) are sharp. By Theorem 6.5(c), there are finitely many planar graphs of diameter 3 with total domination number more than 7. It is entirely possible that there are no such graphs.

Chapter 7
Total Domination and Forbidden Cycles

7.1 Introduction

Recall that Problem 5.1 in Sect. 5.1 is to find for a connected graph G with minimum degree $\delta \geq 1$ and order n an upper bound $f(\delta,n)$ on $\gamma_t(G)$ in terms of δ and n. In this section we consider the following related structural problems.

Problem 7.1. For a given value of $\delta \geq 2$, determine whether the absence of any specified cycle guarantees that the upper bound $f(\delta,n)$ on $\gamma_t(G)$ defined in Problem 5.1 can be lowered.

Problem 7.2. For a graph G of order n, minimum degree $\delta \geq 2$, and girth $g \geq 3$, find a sharp upper bound on $\gamma_t(G)$ in terms of δ, g, and n.

7.2 Forbidden 6-Cycles

Every graph in the family $\mathscr{F}$ of extremal graphs of large order that achieve equality in the bound of Theorem 5.6 has induced 6-cycles. It is therefore a natural question to ask whether this upper bound of $4n/7$ can be improved if we restrict our attention to connected graphs that are C_6-free. For this purpose, the authors in [134] prove a much more general result.

For vertex disjoint subsets X and Y of a graph G, an (X,Y)-total dominating set, abbreviated (X,Y)-TD-set, of G is defined in [134] to be a set S of vertices of G such that $X \cup Y \subseteq S$ and $V(G) \setminus Y \subseteq N(S)$. The *$(X,Y)$-total domination number* of G, denoted by $\gamma_t(G;X,Y)$, is the minimum cardinality of an (X,Y)-TD-set. An (X,Y)-TD-set of G of cardinality $\gamma_t(G;X,Y)$ is called a $\gamma_t(G;X,Y)$-set.

Hence for vertex disjoint subsets X and Y of a graph G, the (X,Y)-total domination number of G is the minimum cardinality of a set $S \subseteq V(G)$ such that

M.A. Henning and A. Yeo, *Total Domination in Graphs*, Springer Monographs in Mathematics, DOI 10.1007/978-1-4614-6525-6_7,

Fig. 7.1 The graph G_7

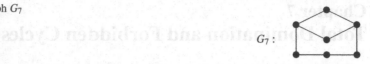

S contains $X \cup Y$ and S totally dominates the set $V(G) \setminus Y$. Note that the $(\emptyset,\emptyset)$-total dominating sets in G are precisely the total dominating sets in G. Thus, $\gamma_t(G) = \gamma_t(G;\emptyset,\emptyset)$.

We remark that the concept of an (X,Y)-TD-set is related to restricted domination in graphs when certain vertices are specified to be in the dominating set. The concept of restricted domination in graphs, where we restrict the dominating sets to contain any given subset of vertices, was introduced by Sanchis in [179] and studied further, for example, in [76, 104]. Restricted total domination in graphs was introduced and studied in [105].

Let G be a graph and let X and Y be disjoint vertex sets in G. A vertex, x, in G is called an (X,Y)-cut-vertex in [134] if the following holds: $G - x$ contains a component which is an induced 6-cycle, C_x, and which does not contain any vertices from X or Y. Furthermore x is adjacent to exactly one vertex on C_x or it is adjacent to exactly two vertices at distance two apart on C_x. Let $\mathrm{bc}(G;X,Y)$ (standing for "bad cut vertex") denote the number of (X,Y)-cut-vertices in G. When $X = Y = \emptyset$, we call an (X,Y)-cut-vertex of G a *bad cut vertex* of G and we denote $\mathrm{bc}(G;X,Y)$ simply by $\mathrm{bc}(G)$. Thus, $\mathrm{bc}(G)$ is the number of bad cut vertices in G.

We also need the following additional terminology defined in [134]. Let G be a graph and let X and Y be disjoint vertex sets in G. Let $\delta_1(G;X,Y)$ denote the number of degree-1 vertices in G that do not belong to Y, and let $\delta_{2,1}(G;X,Y)$ denote the number of degree-2 vertices in G that do not belong to $X \cup Y$ and are adjacent to a degree-1 vertex in G that does not belong to $X \cup Y$.

Before stating the main result in [134], we define a family $\mathscr{C}$ of graphs defined in [134] as follows. Let G_7 be the graph shown in Fig. 7.1.

By *contracting* two vertices x and y in G, we mean replacing the vertices x and y by a new vertex v_{xy} and joining v_{xy} to all vertices in $V(G) \setminus \{x,y\}$ that were adjacent to x or y in G. Let $\mathscr{C}_3$ be a set of graphs only containing one element, namely, the 3-cycle C_3. Similarly let $\mathscr{C}_5 = \{C_5\}$ and $\mathscr{C}_6 = \{C_6\}$. For notational convenience, let $\mathscr{C}_4 = \emptyset$. For every $i > 6$, define $\mathscr{C}_i$ as follows.

For every $i > 6$, a graph G_i belongs to $\mathscr{C}_i$ if and only if $\delta(G_i) \geq 2$ and there is a path v_1, u_1, u_2, u_3, v_3 in G_i on five vertices such that $d(u_1) = d(u_2) = d(u_3) = 2$ in G_i and the graph obtained by contracting v_1 and v_3 and deleting $\{u_1,u_2,u_3\}$ belongs to $\mathscr{C}_{i-4}$. Note that it is possible that v_1v_3 is an edge of G_i.

The family $\mathscr{C}_9$, for example, is shown in Fig. 7.2. Note that $\{H'_{10}, H_{10}\} \subset \mathscr{C}_{10}$, where H'_{10} and H_{10} are the two graphs shown in Fig. 5.1a and b, respectively.

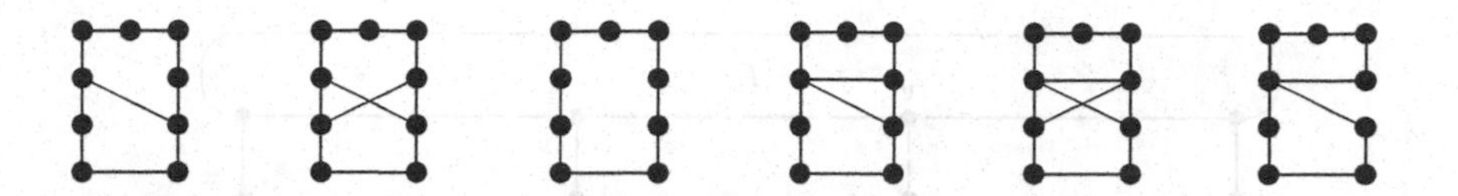

Fig. 7.2 The six graphs in the family $\mathscr{C}_9$

Definition 7.1. Let $\mathscr{C} = \mathscr{C}_3 \cup \mathscr{C}_5 \cup \mathscr{C}_6 \cup \mathscr{C}_7 \cup \mathscr{C}_9 \cup \mathscr{C}_{10} \cup \mathscr{C}_{14} \cup \mathscr{C}_{18} \cup \{K_2, G_7\}$.

In [134] the following lemmas are proved.

Lemma 7.1 ([134]). *If $G \in \mathscr{C}$, then $\gamma_t(G) = \lfloor n(G)/2 \rfloor + 1$.*

Lemma 7.2 ([134]). *If $G \in \mathscr{C} \setminus \{K_2, C_3, C_5, C_6, C_{10}, H_{10}, H'_{10}\}$, then $\gamma_t(G) \le 4n(G)/7$.*

We are now in a position to state the main result in [134]. We remark that the proof of Theorem 7.1 relies on an interplay between total domination in graphs and transversals in hypergraphs.

Theorem 7.1 ([134]). *If G is a connected graph of order at least 2, and if X and Y are disjoint vertex sets in G, then either $X = Y = \emptyset$ and $G \in \mathscr{C}$ or*

$$11\gamma_t(G;X,Y) \le 6|V(G)| + 8|X| + 5|Y| + 2\text{bc}(G;X,Y) \\ +2\delta_1(G;X,Y) + 2\delta_{2,1}(G;X,Y).$$

Note that if G is a graph with $\delta(G) \ge 2$, and X and Y are disjoint vertex sets in G, then $\delta_1(G;X,Y) = \delta_{2,1}(G;X,Y) = 0$. Hence setting $X = Y = \emptyset$ in Theorem 7.1, we have the following consequence of Theorem 7.1.

Corollary 7.1. *If G is a connected graph with $\delta(G) \ge 2$, then either $G \in \mathscr{C}$ or*

$$\gamma_t(G) \le \frac{6}{11}|V(G)| + \frac{2}{11}\text{bc}(G).$$

We remark that if G is a graph of order n, then $\text{bc}(G) \le n/7$. Hence Theorem 5.5 is an immediate consequence of Lemma 7.2 and Corollary 7.1.

If G contains no induced 6-cycle, then $\text{bc}(G) = 0$. Hence since every graph in the family $\mathscr{C}$ has order at most 18, by Corollary 7.1, we have the following main result of this section.

Theorem 7.2 ([134]). *If G is a connected graph of order $n \ge 19$ with $\delta(G) \ge 2$ that has no induced 6-cycle, then $\gamma_t(G) \le 6n/11$.*

To illustrate the sharpness of Theorem 7.2, let $\mathscr{H}_{6/11}$ be the family of all graphs that can be obtained from a connected C_6-free graph H of order at least 2 as follows: For each vertex v of H, add a 10-cycle and join v to exactly one vertex of this cycle. Each graph $G \in \mathscr{H}_{6/11}$ is a connected C_6-free graph of order n with $\gamma_t(G) = 6n/11$. A graph G in the family $\mathscr{H}_{6/11}$ is illustrated in Fig. 7.3 (here the graph H is a 4-cycle).

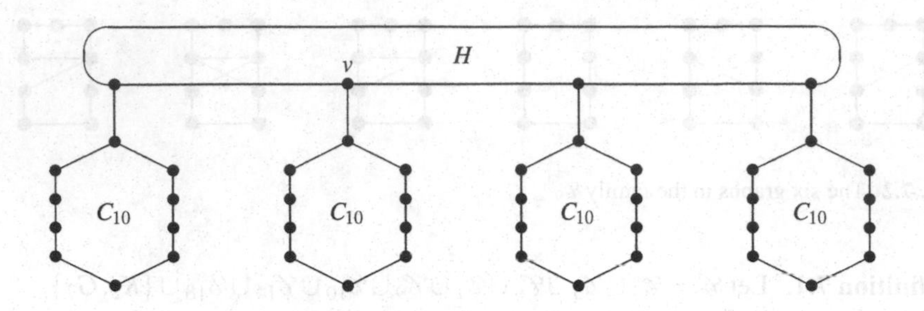

Fig. 7.3 A graph in the family $\mathscr{H}_{6/11}$

7.3 Total Domination and Girth

In this section we consider Problem 7.2 defined earlier in this chapter. The first such result providing an upper bound on $\gamma_t(G)$ in terms of its girth g and order n appeared in [88].

Theorem 7.3 ([88]). *If $G \neq C_n$ is a connected graph with order n, girth $g \geq 7$, and $\delta(G) \geq 2$, then $\gamma_t(G) \leq (4n-g)/6$ unless $G = G_1$, where G_1 is the graph shown in Fig. 7.4, in which case $\gamma_t(G) = 8 = (4n+2-g)/6$.*

The following result both allows smaller girth and gives better bounds for all sufficiently large graphs of any given girth $g \geq 7$. The proof of Theorem 7.4 is once again an interplay between total domination in graphs and transversals in hypergraphs.

Theorem 7.4 ([132]). *If G is a graph of order n, minimum degree at least two, and girth $g \geq 3$, then*

$$\gamma_t(G) \leq \left(\frac{1}{2} + \frac{1}{g}\right) n.$$

By Observation 2.9, the total domination number of a cycle C_n on $n \geq 3$ vertices is given by $\gamma_t(C_n) = \lfloor n/2 \rfloor + \lceil n/4 \rceil - \lfloor n/4 \rfloor$. In particular, if $n \equiv 2 \pmod 4$ and $G = C_n$, then G has order n, girth $g = n$, and

$$\gamma_t(G) = \frac{n+2}{2} = \left(\frac{1}{2} + \frac{1}{g}\right) n.$$

Hence the bound in Theorem 7.4 is sharp for cycles of length congruent to two modulo four.

We remark that by Theorem 7.4, if G is a graph of order n with $\delta(G) \geq 2$ and girth $g \geq 3$, then

Fig. 7.4 The graph G_1

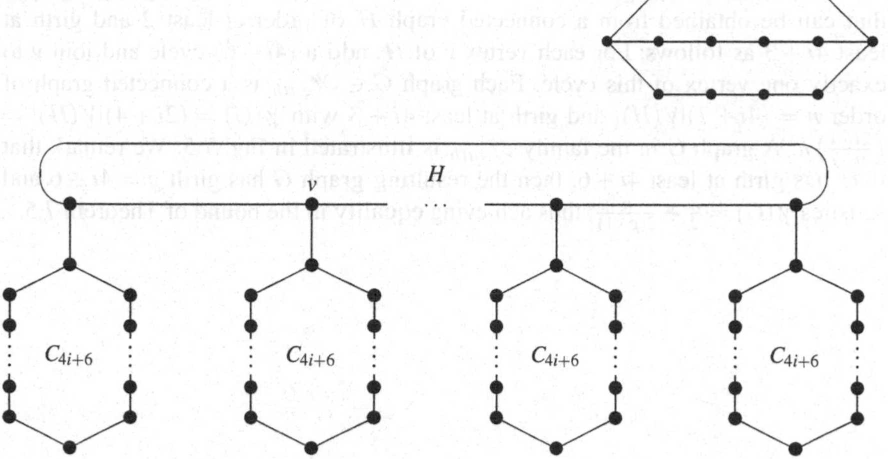

Fig. 7.5 A graph in the family $\mathscr{H}_{girth}$

$$\gamma_t(G) \le \left(\frac{1}{2} + \frac{1}{g}\right) n = \frac{4n - (n - 6n/g)}{6}.$$

Therefore Theorem 7.4 improves on the bound of Theorem 7.3 when $n - 6n/g > g$, i.e., when $n > g/(1 - 6/g) = g^2/\ (g-6)$. We also remark that Theorem 7.4 improves on the result of Theorem 5.5 for large enough girth, namely, for girth $g > 14$.

The upper bounds in both Theorems 7.3 and 7.4 on the total domination number of a graph with minimum degree at least 2 in terms of its order and girth are improved in [137].

Theorem 7.5 ([137]). *If G is a connected graph order n, girth $g \ge 3$ with $\delta(G) \ge 2$, then*

$$\gamma_t(G) \le \frac{n}{2} + \max\left(1, \frac{n}{2(g+1)}\right),$$

and this bound is sharp.

As $n \ge g$, we have that $(\frac{1}{2} + \frac{1}{g})n \ge \frac{n}{2} + 1$, and therefore Theorem 7.5 is a stronger result than Theorem 7.4.

We remark that Theorem 7.5 is a consequence of a much more general result proven in [137], as is the following result.

Theorem 7.6 ([137]). *For $i \ge 1$ an integer, if G is a connected graph of order $n > 8i + 10$ and girth $g \ge 4i + 3$ with $\delta(G) \ge 2$, then*

$$\gamma_t(G) \le \left(\frac{2i+4}{4i+7}\right) n.$$

To illustrate the sharpness of Theorem 7.6, let $\mathscr{H}_{girth}$ be the family of all graphs that can be obtained from a connected graph H of order at least 2 and girth at least $4i+3$ as follows: For each vertex v of H, add a $(4i+6)$-cycle and join v to exactly one vertex of this cycle. Each graph $G \in \mathscr{H}_{girth}$ is a connected graph of order $n = (4i+7)|V(H)|$ and girth at least $4i+3$ with $\gamma_t(G) = (2i+4)|V(H)| = \left(\frac{2i+4}{4i+7}\right) n$. A graph G in the family $\mathscr{H}_{girth}$ is illustrated in Fig. 7.5. We remark that if H has girth at least $4i+6$, then the resulting graph G has girth $g = 4i+6$ and satisfies $\gamma_t(G) = \frac{n}{2} + \frac{n}{2(g+1)}$, thus achieving equality in the bound of Theorem 7.5.

Chapter 8
Relating the Size and Total Domination Number

8.1 Introduction

A classical result of Vizing [210] relates the size and the domination number of a graph of given order. In this chapter we relate the size and the total domination number of a graph of given order.

8.2 Relating the Size and Total Domination Number

Dankelmann et al. [43] established the following Vizing-like relation between the size and the total domination number of a graph of given order. They prove that:

Theorem 8.1 ([43]). *If G is a graph without isolated vertices of order n, size m, and total domination number γ_t, then*

$$m \le \begin{cases} \binom{n-\gamma_t+2}{2} + \frac{\gamma_t}{2} - 1 & \text{if } \gamma_t \text{ is even,} \\ \binom{n-\gamma_t+1}{2} + \frac{\gamma_t}{2} + \frac{1}{2} & \text{if } \gamma_t \text{ is odd.} \end{cases}$$

In order to see that the bound of Theorem 8.1 is sharp, consider the graphs $G(n,k)$ and $H(n,k)$ defined as follows. For $k \ge 2$ even and $n \ge k$, let $G(n,k)$ be the disjoint union of K_{n-k+2} and $(k-2)/2$ copies of K_2. Then, $G(n,k)$ is a graph without isolated vertices of order n, total domination number $\gamma_t = k$, and size

$$m = \binom{n-k+2}{2} + \frac{k-2}{2} = \binom{n-\gamma_t+2}{2} + \frac{\gamma_t}{2} - 1.$$

For $k \ge 3$ odd and $n \ge k+2$, let $H(n,k)$ be the graph obtained from $G(n-2, k-1)$ by subdividing one edge of the component isomorphic to K_{n-k+1} twice. Then,

M.A. Henning and A. Yeo, *Total Domination in Graphs*, Springer Monographs in Mathematics, DOI 10.1007/978-1-4614-6525-6_8,

$H(n,k)$ is a graph without isolated vertices of order n, total domination number $\gamma_t = k$, and size

$$m = \binom{n-k+1}{2} + 2 + \frac{k-3}{2} = \binom{n-\gamma_t+1}{2} + \frac{\gamma_t}{2} + \frac{1}{2}.$$

Restricting their attention to bipartite graphs, the authors in [43] improved the bound of Theorem 8.1 as follows.

Theorem 8.2 ([43]). *If G is a bipartite graph without isolated vertices of order n, size m, and total domination number γ_t, then $m \le \frac{1}{4}((n-\gamma_t)(n-\gamma_t+6)+2\gamma_t)$, and this bound is sharp for $\gamma_t \ge 4$ even.*

We show next that the bound of Theorem 8.2 is sharp for $\gamma_t \ge 4$ even. For $x \ge 5$, the authors in [43] let $H(x)$ denote the balanced or almost balanced complete bipartite graph on x vertices with a minimum edge cover removed, whereby an *edge cover* we mean a set of edges such that every vertex is incident with at least one of the edges. Thus, $H(x)$ is obtained from a complete bipartite graph $K_{\lfloor \frac{x}{2} \rfloor, \lceil \frac{x}{2} \rceil}$ by removing a minimum edge cover. For $k \ge 4$ and $n \ge k+1$, we now let $F(n,k)$ be the disjoint union of $H(n-k+4)$ and $(k-4)/2$ copies of K_2. Then, $F(n,k)$ is a bipartite graph without isolated vertices of order n, total domination number $\gamma_t = k$, and size

$$\begin{aligned} m &= \left\lfloor \frac{n-k+4}{2} \right\rfloor \left\lceil \frac{n-k+4}{2} \right\rceil - \left\lceil \frac{n-k+4}{2} \right\rceil + \frac{k-4}{2} \\ &= \left\lfloor \frac{1}{4}((n-k+4)(n-k+2)+(2k-8)) \right\rfloor \\ &= \left\lfloor \frac{1}{4}((n-k)(n-k+6)+2k) \right\rfloor \\ &= \left\lfloor \frac{1}{4}((n-\gamma_t)(n-\gamma_t+6)+2\gamma_t) \right\rfloor. \end{aligned}$$

When $\gamma_t \ge 3$ is odd, it is unlikely that the bound of Theorem 8.2 is sharp and the authors in [43] pose the conjecture (see Conjecture 18.8 in Chap. 18) that in this case

$$m \le \left\lfloor \frac{1}{4}((n-\gamma_t)(n-\gamma_t+4)+2\gamma_t-2) \right\rfloor.$$

As observed above, the bounds in Theorem 8.1 are sharp, but the edges of the graphs presented in [43] that achieve equality are unevenly distributed, i.e., $\delta(G)$ and $\Delta(G)$ differ greatly ($\delta(G) = 1$ while $\Delta(G) = n - \gamma_t + 1$). If G is connected and $\gamma_t(G) \ge 5$, then Sanchis [181] improved the bound of Theorem 8.1 slightly. For $n \ge 1$, let $F_n = \frac{n}{2}K_2$ if n is even and let $F_n = K_1 \cup (\frac{n-1}{2})K_2$ if n is odd.

Theorem 8.3 ([181]). *If G is a connected graph of order n, size m, and total domination number $\gamma_t \ge 5$, then $m \le \binom{n-\gamma_t+1}{2} + \lfloor \frac{\gamma_t}{2} \rfloor$. If G achieves equality in this bound, then it has one of the following forms:*

1. *G is obtained from $K_{n-\gamma_t} \cup F_{\gamma_t}$ by adding edges between the clique and the graph F_{γ_t} in such a way that each vertex in the clique is adjacent to exactly one vertex in F_{γ_t} and each component of F_{γ_t} has at least one vertex adjacent to a vertex in the clique.*
2. *For $\gamma_t = 5$ and $n \geq 9$, G is obtained from $K_{n-7} \cup P_3 \cup P_4$ by joining every vertex in the clique to both ends of the P_4 and to exactly one end of the P_3 in such a way that each end of the P_3 is adjacent to at least one vertex in the clique.*
3. *For $\gamma_t = 5$ and $n \geq 9$, G is obtained $K_{n-6} \cup F_6$ by letting S be a maximum independent set in F_6 and joining every vertex in the clique to exactly two vertices of S in such a way that each vertex in S is adjacent to at least one vertex in the clique.*

Note that the graphs achieving equality in the bound of Theorem 8.3 have large maximum degree, namely, $\Delta(G) = n - \gamma_t(G)$. In [106, 184] the square dependence on n and γ_t in Theorems 8.1 and 8.3 is improved into a linear dependence on n, γ_t and Δ by demanding a more even distribution of the edges by restricting the maximum degree Δ. Hence a linear Vizing-like relation is established relating the size of a graph and its order, total domination number, and maximum degree.

Theorem 8.4 ([106, 184]). *Let G be a graph each component of which has order at least 3, and let G have order n, size m, total domination number γ_t, and maximum degree $\Delta(G)$. If $\Delta(G) \geq 3$, then $m \leq \Delta(G)(n - \gamma_t)$ and if $\Delta(G) = 2$, then $m \leq 3(n - \gamma_t)$.*

If G is a graph of order n and size m, then $m \leq \Delta(G)n/2$, and so Theorem 8.4 clearly holds when $\gamma_t \leq n/2$. Recall that by Theorem 5.7 in Sect. 5.5, if G is a graph of order n with $\delta(G) \geq 3$, then $\gamma_t(G) \leq n/2$. Therefore Theorem 8.4 holds when $\delta(G) \geq 3$. A characterization of the graphs achieving equality in Theorem 8.4 will be given below.

Theorem 8.5 ([106, 184]). *Let G be a graph with maximum degree at most 3 and with each component of order at least 3. If G has order n and size m, then $\gamma_t(G) \leq n - m/3$.*

The extremal graphs achieving equality in the upper bound in Theorem 8.4 are characterized in [109]; that is, the graphs G satisfying the statement of Theorem 8.4 such that $m = \Delta(n - \gamma_t)$ are characterized.

Let $\mathscr{G}_{\text{cubic}} = \mathscr{G} \cup \mathscr{H} \cup \{G_{16}\}$, where G_{16} is the generalized Petersen graph of order 16 shown in Fig. 1.2 and $\mathscr{G}$ and $\mathscr{H}$ are the two infinite families described in Sect. 5.5. We note that each graph in the family $\mathscr{G}_{\text{cubic}}$ is a cubic graph.

Let $\mathscr{G}_{\delta=1}$ denote the family of all 2-coronas of cycles; that is, $\mathscr{G}_{\delta=1}$ is the family of graphs $H \circ P_2$, where H is a cycle C_k on $k \geq 3$ vertices. We note that each graph in the family $\mathscr{G}_{\delta=1}$ has minimum degree $\delta = 1$. The graph $C_4 \circ P_2 \in \mathscr{G}_{\delta=1}$, for example, is shown in Fig. 8.1.

The authors in [109] define a family $\mathscr{G}_{\delta=2}$ of graphs with minimum degree two. For this purpose, for each $i \geq 0$, they define the family $\mathscr{G}_{4i+3}$ to consist of precisely one graph F_i, namely, the graph $F_i = C_3$ when $i = 0$ and, for $i \geq 1$, the graph F_i

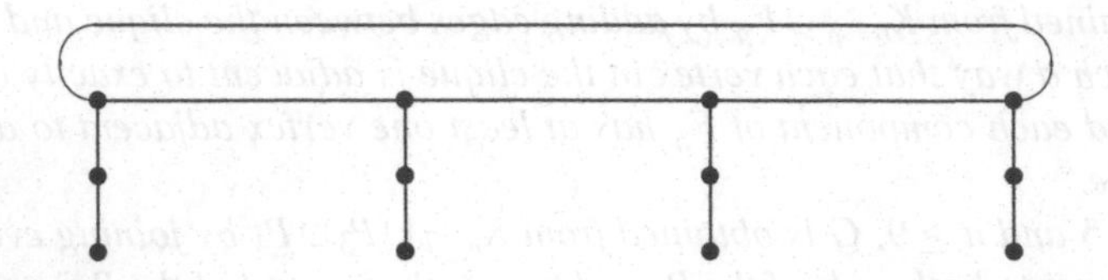

Fig. 8.1 The graph $C_4 \circ P_2 \in \mathscr{G}_{\delta=1}$

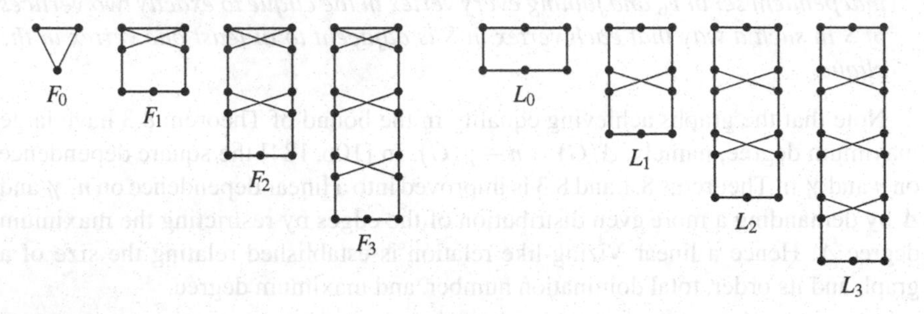

Fig. 8.2 The graphs F_0, F_1, F_2, F_3 and L_0, L_1, L_2, L_3

which is obtained from the graph G_i defined in Sect. 5.5 by subdividing the edge a_1c_1 three times. Further for each $i \geq 0$ they define the family $\mathscr{G}_{4i+2}$ to consist of precisely one graph L_i, namely, the graph $L_i = C_6$ when $i = 0$ and, for $i \geq 1$, the graph L_i which is obtained from the graph G_i defined in Sect. 5.5 by subdividing the edge a_1c_1 three times and subdividing the edge b_kd_k three times. The graphs F_0, F_1, F_2, F_3 and L_0, L_1, L_2, L_3, for example, are shown in Fig. 8.2.

Let $\mathscr{F} = \{F_k \mid k \geq 0\}$ and let $\mathscr{L} = \{L_k \mid k \geq 0\}$. Let $\mathscr{G}_{\delta=2} = \mathscr{F} \cup \mathscr{L}$. We note that each graph in the family $\mathscr{G}_{\delta=2}$ has minimum degree $\delta = 2$.

We are now in a position to present the characterization in [109] of the extremal graphs achieving equality in the upper bound in Theorem 8.4.

Theorem 8.6 ([109]). *Let G be a connected graph of order n, size m, total domination number γ_t, and maximum degree $\Delta(G)$ with each component of G of order at least* 3 *and let $\Delta = \max\{3, \Delta(G)\}$ Then, $m \leq \Delta(n - \gamma_t)$, with equality if and only if $G \in \mathscr{G}_{\delta=1} \cup \mathscr{G}_{\delta=2} \cup \mathscr{G}_{\text{cubic}}$.*

The extremal graphs in Theorem 8.4 also achieve equality in Theorem 8.5. However we remark that all the extremal graphs G have $\Delta(G) \leq 3$. Hence as a consequence of Theorem 8.4, we have the following result.

Corollary 8.7 ([109]). *Let G be a graph each component of which has order at least* 3*, and let G have order n, size m, total domination number γ_t, and maximum degree Δ with $\Delta \geq 4$. Then, $m < \Delta(n - \gamma_t)$.*

It was conjectured in [106] that the bound in the statement of Theorem 8.4 should be $m \leq \frac{1}{2}(\Delta + 3)(n - \gamma_t)$, which is equivalent to conjecturing that $r_\Delta \leq 3$ below.

Problem 8.1 ([106,207]). For each $\Delta \geq 3$, find the smallest value r_Δ, such that any connected graph G of order $n \geq 3$, size m, total domination number γ_t, and maximum degree at most Δ has $m \leq \frac{1}{2}(\Delta + r_\Delta)(n - \gamma_t)$.

Note that Theorem 8.4 implies that $r_\Delta \leq \Delta$. It was shown in Yeo [207] that the conjecture $r_\Delta \leq 3$ is false when $\Delta \geq 1.07 \times 10^{13}$, as the following was proven using results on both transversals in hypergraphs and graph theory.

Theorem 8.8 ([207]). *If r_Δ is defined as in Problem 8.1, then $0.1\ln(\Delta) \leq r_\Delta \leq 2\sqrt{\Delta}$, for all $\Delta \geq 3$.*

One can also write the results as an upper bound on the total domination number and in fact the following was proven in [207].

Theorem 8.9 ([207]). *If G is a connected graph of order $n \geq 3$, size m, total domination number γ_t, and maximum degree at most Δ, then*

$$\gamma_t \leq n - \left(\frac{2}{\Delta + 2\sqrt{\Delta}}\right) m.$$

Furthermore if $\Delta = 4$, then this can be strengthened to $\gamma_t \leq n - 6m/21$.

Chapter 9
Total Domination in Claw-Free Graphs

9.1 Introduction

In this chapter, we impose the structural restriction of claw-freeness on a graph and investigate upper bounds on the total domination number of such graphs.

9.2 Minimum Degree One

If we restrict G to be a connected claw-free graph, then the upper bound of Theorem 5.2 in Chap. 5 cannot be improved since the 2-corona of a complete graph is claw-free and has total domination number two-thirds its order. Hence it is only interesting to consider claw-free graphs with minimum degree at least two.

9.3 Minimum Degree Two

Every graph in the family $\mathscr{F}$ of extremal graphs of large order that achieve equality in the bound of Theorem 5.6 (see Sect. 5) contains a claw. It is therefore a natural question to ask whether the upper bound of Theorem 5.6 can be improved if we restrict G to be a connected claw-free graph. For this purpose, we construct an infinite family $\mathscr{G}^*$ of connected, claw-free graphs G of order n satisfying $\gamma_t(G) = (n+1)/2$. Let $G_1, G_2, \ldots, G_7$ be the seven graphs shown in Fig. 9.1.

We define an *elementary* 4-*subdivision* of a nonempty graph G as a graph obtained from G by subdividing some edge four times. A 4-*subdivision* of G is a graph obtained from G by a sequence of zero or more elementary 4-subdivisions. We define a *good edge* of a graph G to be an edge uv in G such that both $N[u]$ and $N[v]$ induce a clique in $G - uv$. Further, we define a *good* 4-*subdivision* of G to be a 4-subdivision of G obtained by a sequence of elementary 4-subdivisions of good

M.A. Henning and A. Yeo, *Total Domination in Graphs*, Springer Monographs in Mathematics, DOI 10.1007/978-1-4614-6525-6_9,

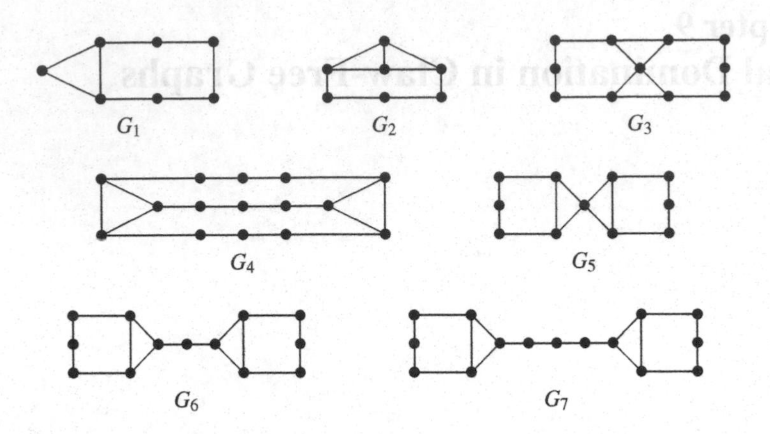

Fig. 9.1 The graphs $G_1, G_2, \ldots, G_7$

edges (at each stage in the resulting graph). For $i = 1, 2, \ldots, 7$, let $\mathscr{G}_i^* = \{G \mid G$ is a good 4-subdivision of $G_i\}$. Let $\mathscr{G}^*$ be the family define by

$$\mathscr{G}^* = \bigcup_{i=1}^{7} \mathscr{G}_i^*.$$

We are now in a position to state the following upper bound on the total domination of a claw-free graph presented in [64].

Theorem 9.1 ([64]). *If G is a connected claw-free graph of order n with $\delta(G) \geq 2$, then either*

(i) $\gamma_t(G) \leq n/2$, *or*
(ii) *G is an odd cycle or $G \in \mathscr{G}^*$, in which case $\gamma_t(G) = (n+1)/2$, or*
(iii) *$G = C_n$ where $n \equiv 2 \pmod 4$, in which case $\gamma_t(G) = (n+2)/2$.*

Corollary 9.2 ([64]). *If G is a connected claw-free graph of order n with $\delta(G) \geq 2$, then $\gamma_t(G) \leq (n+2)/2$ with equality if and only if G is a cycle of length congruent to 2 modulo 4.*

9.4 Cubic Graphs

Every graph in the two families $\mathscr{G}$ and $\mathscr{H}$ of extremal graphs that achieve equality in the bound of Theorem 5.10 (see Sect. 5.5), as well as the generalized Petersen graph G_{16} shown in Fig. 1.2, is a cubic graph. However all these extremal graphs, except for K_4 and the cubic graph G_1 shown in Fig. 9.2, contain a claw. It is

Fig. 9.2 A claw-free cubic graph G_1 with $\gamma_t(G_1) = n/2$

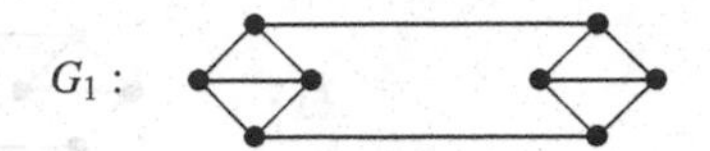

therefore a natural question to ask whether the upper bound of Theorem 5.7 can be improved if we restrict G to be a connected cubic claw-free graph of order at least ten. The connected claw-free cubic graphs achieving equality in Theorem 5.7 are characterized in [63].

Theorem 9.3 ([63]). *If G is a connected claw-free cubic graph of order n, then $\gamma_t(G) \leq n/2$ with equality if and only if $G = K_4$ or $G = G_1$ where G_1 is the graph shown in Fig. 9.2.*

We remark that if G is a connected graph of order n with minimum degree at least 3 that is not a cubic graph, then by Theorem 5.10, $\gamma_t(G) < n/2$. Hence, Theorem 9.3 also holds for connected claw-free graphs of order n with minimum degree at least 3. The result of Theorem 9.3 can be strengthened as follows.

Theorem 9.4 ([65]). *If G is a connected claw-free cubic graph of order $n \geq 6$, then either $G = G_1$, where G_1 is the graph of order 8 shown in Fig. 9.2, or $\gamma_t(G) \leq 5n/11$.*

The authors in [65] posed the conjecture that the upper bound in Theorem 9.4 can be improved from five-elevenths its order to four-ninths its order. This conjecture was solved in [188].

Theorem 9.5 ([188]). *If G is a connected claw-free cubic graph of order $n \geq 10$, then $\gamma_t(G) \leq 4n/9$.*

The following theorem by Lichiardopol [162] characterized the extremal graphs achieving equality in the bound of Theorem 9.5, where the two connected claw-free cubic graphs F_1 and F_2 are shown in Fig. 9.3.

Theorem 9.6 ([162]). *If G is a connected claw-free cubic graph of order $n \geq 10$ satisfying $\gamma_t(G) = 4n/9$, then $G \in \{F_1, F_2\}$.*

Lichiardopol [162] improves the upper bound in Theorem 9.5 for connected claw-free cubic graphs of order at least 20.

Theorem 9.7 ([162]). *If G is a connected claw-free cubic graph of order $n \geq 20$, then $\gamma_t(G) < 10n/23$.*

We remark that the upper bound in Theorem 9.7 is not achieved and that it remains an open problem to determine a sharp upper bound on the total domination of a connected claw-free cubic graph of large order. We refer the reader to the open problem stated in Sect. 18.12.

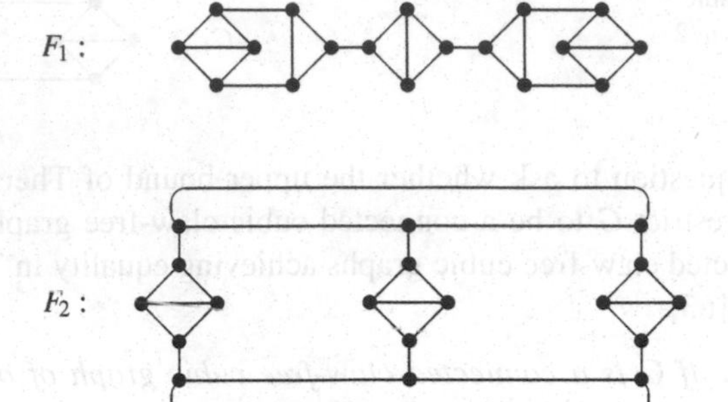

Fig. 9.3 The claw-free cubic graphs F_1 and F_2

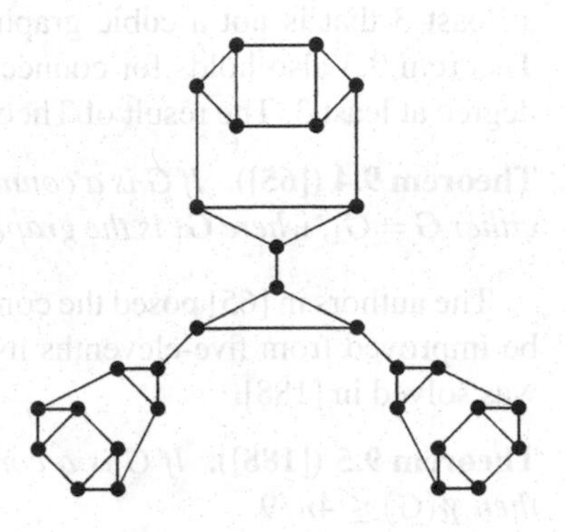

Fig. 9.4 A $(K_{1,3}, K_4 - e)$-free cubic G with $\gamma_t(G) = 2n/5$

If we further restrict our claw-free graph to be diamond free, then the upper bound on the total domination number of G in Theorems 9.4 and 9.7 decreases to $2n/5$.

Theorem 9.8 ([65]). *If G is a connected $(K_{1,3}, K_4 - e)$-free cubic graph of order $n \geq 6$, then $\gamma_t(G) \leq 2n/5$ with equality if and only if G is the graph shown in Fig. 9.4.*

As a consequence of Theorem 9.4 we observe that if G is a connected $(K_{1,3}, K_4 - e)$-free cubic graph of order $n > 30$, then $\gamma_t(G) < 2n/5$. However it remains an open problem to determine a sharp upper bound on the total domination of a connected $(K_{1,3}, K_4 - e)$-free cubic graph of large order.

If we further forbid induced 4-cycles, then the upper bound on the total domination number of G in Theorem 9.8 decreases from two-fifths its order to three-eighths its order.

Theorem 9.9 ([65]). *If G is a connected $(K_{1,3}, K_4 - e, C_4)$-free cubic graph of order $n \geq 6$, then $\gamma_t(G) \leq 3n/8$ with equality if and only if G is the graph shown in Fig. 9.5.*

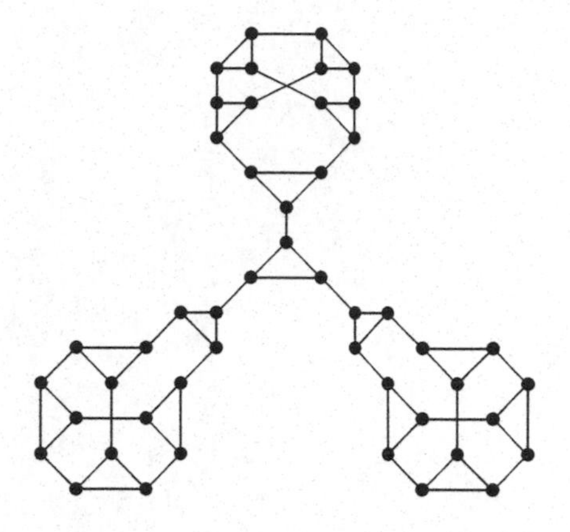

Fig. 9.5 A $(K_{1,3}, K_4 - e, C_4)$-free cubic G with $\gamma_t(G) = 3n/8$

As a consequence of Theorem 9.9 we observe that if G is a connected $(K_{1,3}, K_4 - e, C_4)$-free cubic graph of order $n > 48$, then $\gamma_t(G) < 3n/8$. However it remains an open problem to determine a sharp upper bound on the total domination of a connected $(K_{1,3}, K_4 - e, C_4)$-free cubic graph of large order.

Chapter 10
Total Domination Number Versus Matching Number

10.1 Introduction

Bollobás and Cockayne [16] proved the following classical result: If $G = (V,E)$ is a graph with no isolated vertex, then there exists a minimum dominating set D of G with the property that $\text{epn}[v,D] \neq \emptyset$ for every vertex $v \in D$; that is, for each $v \in D$, there is a vertex $v' \in V \setminus D$ that is adjacent to v but to no other vertex of D in G. As an immediate consequence of this property of minimum dominating sets in graphs, we have that the matching number of a graph G with no isolated vertex is at least its domination number; that is, $\gamma(G) \le \alpha'(G)$. Since $\gamma(G) \le \gamma_t(G)$ for all graphs G with no isolated vertex, it is natural to ask the following questions: Is it true that $\gamma_t(G) \le \alpha'(G)$ for every graph G with sufficiently large minimum degree? For what graph classes $\mathscr{G}'$ is it true that $\gamma_t(G) \le \alpha'(G)$ holds for all graphs $G \in \mathscr{G}'$?

10.2 Relating the Total Domination and Matching Numbers

In general the matching number and total domination number of a graph are incomparable, even for arbitrarily large, but fixed (with respect to the order of the graph), minimum degree. First we consider the case when the minimum degree is small, and prove the following result.

Theorem 10.1 ([113]). *For every integer $k \ge 0$ and for $\delta \in \{1,2\}$, there exists graphs G and H with $\delta(G) = \delta(H) = \delta$ satisfying $\gamma_t(G) - \alpha'(G) > k$ and $\alpha'(H) - \gamma_t(H) > k$.*

Proof. Suppose first that $\delta = 1$. In this case, we let G be the 2-corona of a path P_{2k+2}, and so $G = P_{2k+2} \circ P_2$, and we let H be obtained from the complete graph K_{2k+5} by adding a new vertex (of degree 1) and joining it to one vertex of the complete graph. Then, $n(G) = 6(k+1)$, $\delta(G) = 1$ and $\alpha'(G) = n(G)/2 = 3(k+1)$. By Theorem 2.7, $\gamma_t(G) = 4(k+1)$, implying that $\gamma_t(G) - \alpha'(G) = k+1$. Moreover, $n(H) = 2(k+3)$,

M.A. Henning and A. Yeo, *Total Domination in Graphs*, Springer Monographs in Mathematics, DOI 10.1007/978-1-4614-6525-6_10,

$\delta(H) = 1$ and $\alpha'(H) = n(H)/2 = k+3$, while $\gamma_t(H) = 2$, implying that $\alpha'(H) - \gamma_t(H) = k+1$.

Suppose next that $\delta = 2$. In this case, let G be obtained from a path P_{2k+2} as follows: For each vertex v of the path, add a 6-cycle and join v to one vertex of this cycle. Further let H be obtained from the complete graph K_{2k+5} by adding a new vertex (of degree 2) and joining it to two vertices of the complete graph. Then, $n(G) = 14(k+1)$, $\delta(G) = 2$ and $\alpha'(G) = n(G)/2 = 7(k+1)$. By Theorem 5.5 and Lemma 5.1 in Sect. 5.4, $\gamma_t(G) = 8(k+1)$, implying that $\gamma_t(G) - \alpha'(G) = k+1$. Moreover, $n(H) = 2(k+3)$, $\delta(H) = 2$ and $\alpha'(H) = n(H)/2 = k+3$, while $\gamma_t(H) = 2$, implying that $\alpha'(H) - \gamma_t(H) = k+1$. □

We are now in a position to prove the following result.

Theorem 10.2 ([113]). *For every integer $\delta \geq 1$, there exists graphs G and H with $\delta(G) = \delta(H) = \delta$ satisfying $\gamma_t(G) > \alpha'(G)$ and $\gamma_t(H) < \alpha'(H)$.*

Proof. When $\delta \in \{1,2\}$, the result follows from the stronger result Theorem 10.1. Hence we may assume that $\delta \geq 3$.

For $n \geq (\delta-1)\delta + 1$, let $G = G_n^\delta$ be the bipartite graph formed by taking as one partite set a set A of n elements, and as the other partite set a set B of all the δ-element subsets of A and joining each element of A to those subsets it is a member of. Then, every vertex in B has degree δ, while every vertex in A has degree $\binom{n-1}{\delta-1}$. Thus, G is a bipartite graph with minimum degree δ and order $n + \binom{n}{\delta}$. Now, $\alpha'(G) \leq \min(|A|,|B|) = |A| = n$. It is easy to find a matching of size n in G, and so $\alpha'(G) = n$. To totally dominate the vertices in B we need at least $n - \delta + 1$ vertices in A, while to totally dominate the vertices in A, we need at least $\lceil |A|/\delta \rceil = \lceil n/\delta \rceil$ vertices in B. Hence, $\gamma_t(G_n^\delta) \geq n - \delta + 1 + \lceil n/\delta \rceil$. It is not difficult to see that there exists $\lceil n/\delta \rceil$ vertices in B which totally dominate all vertices in A so these vertices together with any $n - \delta + 1$ vertices in A imply that $\gamma_t(G) = n - \delta + 1 + \lceil n/\delta \rceil$. Thus since $n \geq (\delta-1)\delta + 1$, $\gamma_t(G) > \alpha'(G)$.

Let H be obtained from the complete graph $K_{2\delta-1}$ by adding a new vertex (of degree δ) and joining it to δ vertices of the complete graph. Then, $\delta(H) = \delta$, $\gamma_t(H) = 2$ while $\alpha'(H) = \delta$, whence $\alpha'(H) > \gamma_t(H)$. □

A *path covering* of a graph G is a collection of vertex disjoint paths of G that partition $V(G)$. The minimum cardinality of a path covering of G is the *path covering number* of G, denoted $\mathrm{pc}(G)$. The following result is proven by DeLaViña, Liu, Pepper, Waller, and West [44].

Theorem 10.3 ([44]). *If G is a connected graph of order $n \geq 2$, then $\gamma_t(G) \leq \alpha'(G) + \mathrm{pc}(G)$.*

That the bound of Theorem 10.3 is sharp, may be seen by as follows: For $m \geq 1$, let G_m be the graph of order $7m$ obtained from a cycle C_m, if $m \geq 3$, or a path P_m, if $m = 1$ or $m = 2$, by identifying each vertex of the cycle or path with the center of a path P_7. Then, $\gamma_t(G_m) = 4m$, $\alpha'(G_m) = 3m$ and $\mathrm{pc}(G_m) = m$.

10.3 Graphs with Total Domination Number at Most the Matching Number

By Theorem 10.2, it is not true that $\gamma_t(G) \le \alpha'(G)$ for all graphs G, even if the minimum degree is arbitrarily large (but fixed with respect to the order of the graph). However since the ends of the edges in a maximum matching in a graph form a TD-set in the graph, we observe that for every graph G with no isolated vertex, we have $\gamma_t(G) \le 2\alpha'(G)$. It would be of interest to determine for which graph classes $\mathscr{G}'$ it is true that $\gamma_t(G) \le \alpha'(G)$ holds for all graphs $G \in \mathscr{G}'$.

10.3.1 Claw-Free Graphs

The authors in [113] investigate the question of whether $\gamma_t(G) \le \alpha'(G)$ is true for the family of claw-free graphs with minimum degree at least three and prove that this is indeed the case.

Theorem 10.4 ([113]). *For every claw-free graph G with $\delta(G) \ge 3$, $\gamma_t(G) \le \alpha'(G)$.*

Note that Theorem 10.4 is not true if we relax the minimum degree condition from minimum degree at least three to minimum degree at least two as can be seen by a cycle C_{4i+2} of length 2 modulo 4, as $\gamma_t(C_{4i+2}) > n(C_{4i+2})/2$. The connected claw-free graphs with minimum degree at least three that achieve equality in the bound of Theorem 10.4 are characterized in [129]. For this purpose, the authors in [129] define a collection $\mathscr{F}_{cf}$ of connected claw-free graphs with minimum degree of three and maximum of degree four that have equal total domination and matching numbers. Let $\mathscr{F}_{cf} = \{F_1, F_2, \ldots, F_{12}\}$ be the collection of twelve graphs shown in Fig. 10.1. Let G_1 be the claw-free cubic graph shown in Fig. 9.2 in Sect. 9.4.

Theorem 10.5 ([129]). *Let G be a connected claw-free graph with $\delta(G) \ge 3$. Then, $\gamma_t(G) = \alpha'(G)$ if and only if $G \in \mathscr{F}_{cf} \cup \{K_4, K_5 - e, K_5, G_1\}$.*

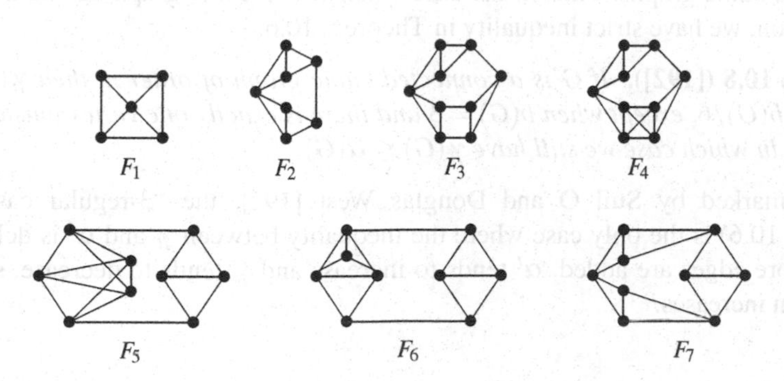

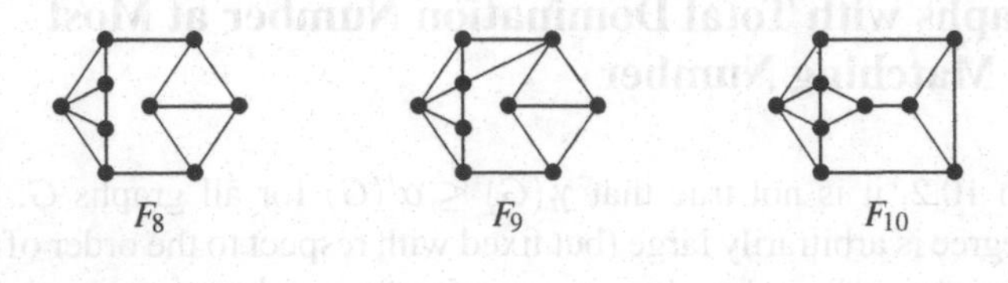

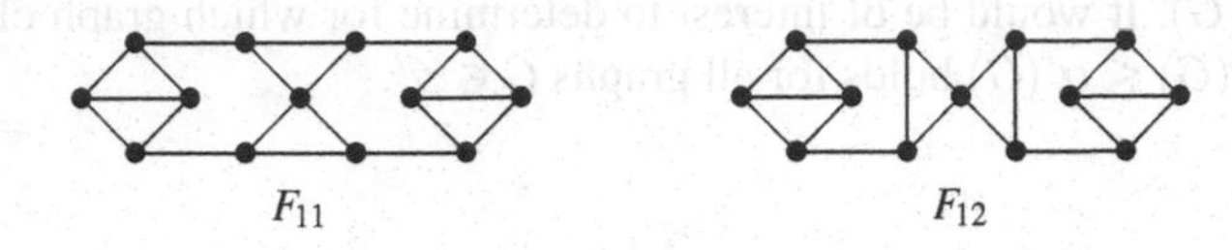

Fig. 10.1 The collection $\mathscr{F}_{cf}$ of twelve graphs

10.3.2 Regular Graphs

The authors in [113] also investigate the question of whether $\gamma_t(G) \leq \alpha'(G)$ is true for the family of k-regular graphs when $k \geq 3$ and answer this question in the affirmative.

Theorem 10.6 ([113]). *For every k-regular graph G with $k \geq 3$, $\gamma_t(G) \leq \alpha'(G)$.*

Recall that a *balloon* in a graph G is defined in Sect. 5.6 to be a maximal 2-edge-connected subgraph incident to exactly one cut-edge of G. Further, $b(G)$ denotes the number of balloons in G. Suil O and Douglas West [192] established the following lower bound on the matching number of a cubic graph in terms of its order and number of balloons.

Theorem 10.7 ([192]). *If G is a connected cubic graph of order n, then $\alpha'(G) \geq n/2 - \lfloor b(G)/3 \rfloor$.*

Theorems 5.14 and 10.7 together improve the inequality $\gamma_t(G) \leq \alpha(G)$ for connected cubic graphs. Thus in the case when G is a cubic graph having at least one balloon, we have strict inequality in Theorem 10.6.

Theorem 10.8 ([192]). *If G is a connected cubic graph of order n, then $\gamma_t(G) \leq \alpha'(G) - b(G)/6$, except when $b(G) = 3$ and there is exactly one vertex outside the balloons, in which case we still have $\gamma_t(G) \leq \alpha(G)$.*

As remarked by Suil O and Douglas West [192], the "3-regular case (in Theorem 10.6) is the only case where the inequality between γ_t and α' is delicate. When more edges are added, α' tends to increase and γ_t tends to decrease, so the separation increases."

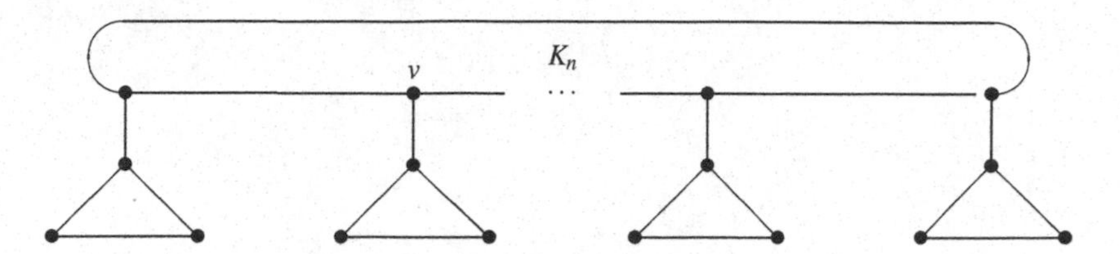

Fig. 10.2 A graph G in the family $\mathscr{F}_{3cycles}$

10.3.3 Graphs with Every Vertex in a Triangle

As a special case of a more general result, the authors in [139] prove the following result.

Theorem 10.9 ([139]). *If G is a graph with no K_3-component and with all vertices of G contained in a triangle, then $\gamma_t(G) \le \alpha'(G)$.*

That the bound in Theorem 10.9 is sharp may be seen as follows. Let $\mathscr{F}_{3cycles}$ be the family of all graphs that can be obtained from a connected graph F in which every vertex belongs to a triangle as follows: For each vertex v of F, add a 3-cycle and join v to one vertex of this cycle. Let G denote the resulting graph. Then, $\gamma_t(G) = \alpha'(G) = |V(G)|/2$. A graph G in the family $\mathscr{F}_{3cycles}$ is illustrated in Fig. 10.2 (here the graph F is a complete graph K_n).

We remark that the requirement in the statement of Theorem 10.9 that all vertices belong to a triangle can be relaxed slightly by allowing all except possibly one vertex to belong to a triangle. However there are connected graphs G of arbitrarily large order with all except four vertices that do not belong to a triangle but with $\gamma_t(G) > \alpha'(G)$. For example, for $k \ge 1$ let G be obtained from k vertex disjoint copies of K_3 by adding a path P_4 on four vertices and joining one of its ends to one vertex from each copy of K_3. Then, $\gamma_t(G) = k+3$ and $\alpha'(G) = k+2$, and so $\gamma_t(G) > \alpha'(G)$. It is not yet known whether the requirement in the statement of Theorem 10.9 can be relaxed by allowing all, except possibly two or three, vertices to belong to a triangle.

Chapter 11
Total Domination Critical Graphs

11.1 Introduction

For many graph parameters, criticality is a fundamental question. Much has been written about those graphs where a parameter (such as connectedness or chromatic number) goes up or down whenever an edge or vertex is removed or added. In this chapter, we consider the same concept for total domination.

11.2 Total Domination Edge-Critical Graphs

Sumner and Blitch [193] began the study of those graphs, called domination edge-critical graphs, where the (ordinary) domination number decreases on the addition of any edge. The study of total domination edge-critical graphs, defined analogously, was initiated by Van der Merwe [202] and Van der Merwe, Haynes, and Mynhardt [203]. A graph G is said to be *total domination edge-critical*, abbreviated $\gamma_t EC$, if $\gamma_t(G+e) < \gamma_t(G)$ for every edge $e \in E(\overline{G}) \neq \emptyset$. Further if G is $\gamma_t EC$ and $\gamma_t(G) = k$, we say that G is $k_t EC$. Thus if G is $k_t EC$, then its total domination number is k and the addition of any edge decreases the total domination number. In particular, we note that by definition the complete graph on at least two vertices is not $2_t EC$. Further since $\gamma_t(G) \geq 2$ for every graph G with no isolated vertex, we note that if G is a $k_t EC$ graph, then $k \geq 3$.

As remarked earlier, if G is a $k_t EC$ graph, then $k \geq 3$. The 5-cycle is a simple example of a $3_t EC$ graph. For $k \geq 3$, it is a difficult problem to characterize $k_t EC$ graphs, even in the special case when $k = 3$. To date the problem to characterize $k_t EC$ graphs remains open for every $k \geq 3$.

M.A. Henning and A. Yeo, *Total Domination in Graphs*, Springer Monographs in Mathematics, DOI 10.1007/978-1-4614-6525-6_11,

11.2.1 Supercritical Graphs

It is shown in [203] that the addition of an edge to a graph can change the total domination number by at most two.

Observation 11.1 ([203]). *If G is a graph with no isolated vertex and $e \in E(\overline{G}) \neq \emptyset$, then $\gamma_t(G) - 2 \leq \gamma_t(G+e) \leq \gamma_t(G)$.*

Total domination edge-critical graphs G with the property that $\gamma_t(G) = k$ and $\gamma_t(G+e) = k-2$ for every edge $e \in E(\overline{G})$ are called *k_t-supercritical graphs*. Since $\gamma_t(G) \geq 2$ for every graph G with no isolated vertex, we note that if G is a k_t-supercritical graph, then $k \geq 4$. The 4_t-supercritical graphs are characterized in [202].

Theorem 11.2 ([202]). *A graph G is 4_t-supercritical if and only if G is the disjoint union of two nontrivial complete graphs.*

The result of Theorem 11.2 can readily be generalized to the following characterization of k_t-supercritical graphs for all $k \geq 4$. Recall that a *clique* is a complete graph and a *nontrivial clique* is a clique on at least two vertices.

Theorem 11.3. *A graph G is k_t-supercritical if and only if $k \geq 4$ is even and G is the disjoint union of $k/2$ nontrivial cliques.*

Proof. Let G be a k_t-supercritical graph. As observed earlier, $k \geq 4$. We first show that G is P_3-free. For the sake of contradiction, let xyz be a path in G where x and z are not adjacent in G. Let S_{xz} be a $\gamma_t(G+xz)$-set. Since G is k_t-supercritical, we have that $|S_{xz}| = \gamma_t(G+xz) = k-2$. However the set $S_{xz} \cup \{y\}$ is a TD-set in G of cardinality less than k, a contradiction. Therefore, G is P_3-free. However the only graphs that are P_3-free are the graphs where every component is a clique. Since G has no isolated vertex, every component of G is therefore a nontrivial clique. Further since $\gamma_t(G) = k$ and two vertices are required to totally dominate the vertices of each clique, the graph G is the disjoint union of $k/2$ nontrivial cliques. Conversely if $k \geq 4$ is even and G is the disjoint union of $k/2$ nontrivial complete graphs, then $\gamma_t(G) = k$ and $\gamma_t(G+e) = k-2$ for every edge $e \in E(\overline{G})$, implying that G is k_t-supercritical. □

11.2.2 General Constructions

In this section, we give a way of constructing a total domination edge-critical graph from two smaller total domination edge-critical graphs. The results in this section can be found in [127]. First we define a family of graphs Q_q, $q \geq 1$, with total domination number $2q$. For $q \geq 1$, let Q_q be the graph constructed as follows. Consider two copies of the path P_{3q} with respective vertex sequences

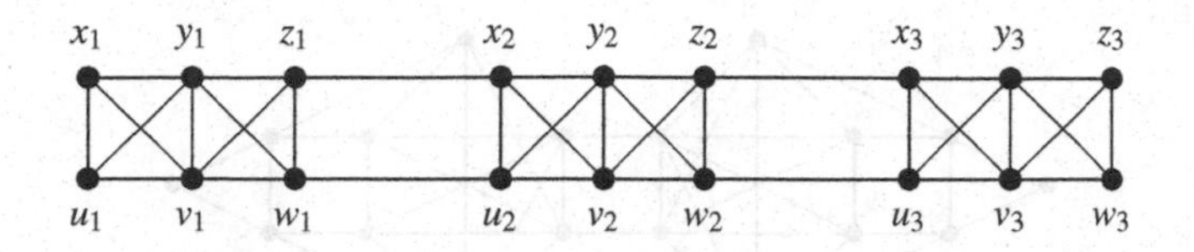

Fig. 11.1 The graph Q_3

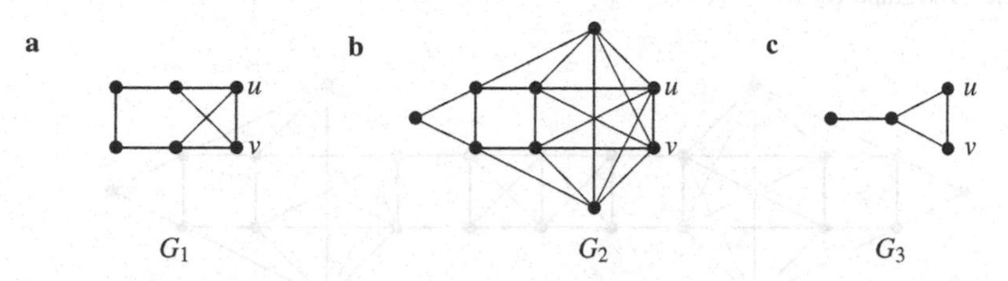

Fig. 11.2 Three γ_t-special graphs

$u_1, v_1, w_1, \ldots, u_q, v_q, w_q$ and $x_1, y_1, z_1, \ldots, x_q, y_q, z_q$. For each $i \in \{1, 2, \ldots, q\}$, join u_i to x_i and y_i; join v_i to x_i, y_i, and z_i; and join w_i to y_i and z_i. The graph Q_3 is illustrated in Fig. 11.1.

Lemma 11.1 ([127]). *For $q \geq 1$, $\gamma_t(Q_q) = 2q$.*

Let $G = (V, E)$ be a graph, and let $X \subset V$. If S is a dominating set in G such that $X \subseteq S$ and $G[S \setminus X]$ contain no isolated vertices, then we call the set S an *almost TD-set of G relative to X*. Thus the set S totally dominates all vertices of G, except possibly for vertices in the set X. The *almost total domination number of G relative to X*, denoted $\gamma_t(G; X)$, is the minimum cardinality of an almost TD-set of G relative to X. If $|X| = 1$, then we simply denote $\gamma_t(G; X)$ by $\gamma_t(G; x)$ where $X = \{x\}$, while if $|X| = 2$, then we denote $\gamma_t(G; X)$ by $\gamma_t(G; x, y)$ where $X = \{x, y\}$.

In [127], a graph G is defined to be *γ_t-special* if G has two distinguished vertices u and v such that $N[u] = N[v]$ and $\gamma_t(G; x) = \gamma_t(G)$ for $x \in \{u, v\}$. The graphs G_1, G_2, and G_3 shown in Fig. 11.2a, b, and c, respectively, are examples of γ_t-special graphs.

If F is a γ_t-special graph with distinguished vertices u and v, and H is a γ_t-special graph with distinguished vertices x and y, then we denote the graph formed from the disjoint union of F and H by adding the edges ux and vy by $F \boxdot H$. The graph $G_2 \boxdot G_2$ is shown in Fig. 11.3, where G_2 is the graph shown in Fig. 11.2b.

If F is a γ_t-special graph with distinguished vertices u and v, and H is a γ_t-special graph with distinguished vertices x and y, then for $q \geq 1$, we denote the graph formed from the disjoint union of F, H, and Q_q by adding the edges $\{ux_1, vu_1, xz_q, yw_q\}$ by $F \boxdot Q_q \boxdot H$. The graph $G_2 \boxdot Q_1 \boxdot G_2$ is shown in Fig. 11.4, where G_2 is the graph shown in Fig. 11.2b.

Lemma 11.2 ([127]). *If F and H are γ_t-special graphs, then for $q \geq 1$, $\gamma_t(F \boxdot Q_q \boxdot H) = \gamma_t(F) + \gamma_t(Q_q) + \gamma_t(H)$.*

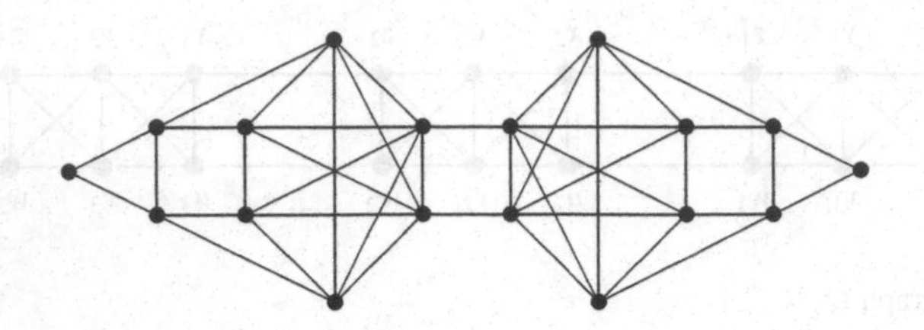

Fig. 11.3 The graph $G_2 \Box G_2$

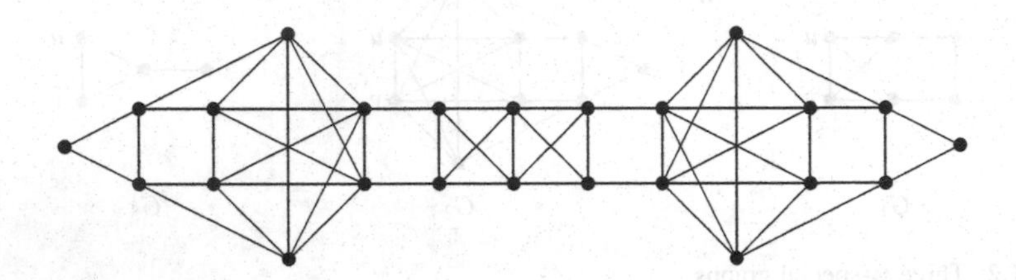

Fig. 11.4 The graph $G_2 \Box Q_1 \Box G_2$

Lemma 11.3 ([127]). *If F and H are γ_t-special graphs, then* $\gamma_t(F \Box H) = \gamma_t(F) + \gamma_t(H)$.

A special family of total domination edge-critical graphs that satisfy certain desirable properties is defined in [127] as follows. A graph G is a *special $\gamma_t EC$ graph* if G is a $\gamma_t EC$ graph that is γ_t-special with distinguished vertices u and v satisfying the following properties where $x \in \{u,v\}$:

(a) There is a $\gamma_t(G)$-set containing x.
(b) $\gamma_t(G-u-v) = \gamma_t(G) - 1$.
(c) For every vertex $w \in V \setminus \{u,v\}$, $\gamma_t(G;x,w) < \gamma_t(G)$ or $\gamma_t(G-w) < \gamma_t(G)$. Furthermore, if $\gamma_t(G;x,w) \geq \gamma_t(G)$ or $\gamma_t(G-w) \geq \gamma_t(G)$, then there is a $\gamma_t(G)$-set containing w.

If G is a special $\gamma_t EC$ graph and $\gamma_t(G) = k$, we say that G is a *special $k_t EC$ graph*. The graphs G_1 and G_2 shown in Fig. 11.2a and b, respectively, are examples of special $3_t EC$ graphs.

Lemma 11.4 ([127]). *If F is a special $j_t EC$ graph and H is a special $k_t EC$ graph, then for $q \geq 1$, $G = F \Box Q_q \Box H$ is a $(k+j+2q)_t EC$ graph.*

Lemma 11.5 ([127]). *If F is a special $j_t EC$ graph and H is a special $k_t EC$ graph, then $F \Box H$ is a $(k+j)_t EC$ graph.*

By Lemma 11.4, the graph $G_2 \Box Q_1 \Box G_2$ shown in Fig. 11.4 is a $8_t EC$ graph. By Lemma 11.5, the graph $G_2 \Box G_2$ shown in Fig. 11.3 is a $6_t EC$ graph.

A graph G is defined in [127] to be an *almost $\gamma_t EC$ graph* if G is γ_t-special with distinguished vertices u and v satisfying the following properties where $x \in \{u,v\}$:

(a) There is a $\gamma_t(G)$-set containing x.
(b) $\gamma_t(G+e;u) < \gamma_t(G)$ or $\gamma_t(G+e;v) < \gamma_t(G)$ for every edge $e \in E(\overline{G})$.
(c) For every vertex $w \in V \setminus \{u,v\}$, $\gamma_t(G-w;v) < \gamma_t(G)$ or $\gamma_t(G;w) < \gamma_t(G)$. Furthermore, if $\gamma_t(G-w;v) \geq \gamma_t(G)$ or $\gamma_t(G;w) \geq \gamma_t(G)$, then there is a $\gamma_t(G)$-set containing w.

The graph G_3 shown in Fig. 11.2c is the simplest example of an almost 2_tEC graph.

Lemma 11.6 ([127]). *If F is special j_tEC graph and H is an almost k_tEC graph, then for $q \geq 1$, $G = F \boxdot Q_q \boxdot H$ is a $(k+j+2q)_tEC$ graph.*

Lemma 11.7 ([127]). *If F is special j_tEC graph and H is an almost k_tEC graph, then $F \boxdot H$ is a $(k+j)_tEC$ graph.*

11.2.3 The Diameter of $\gamma_t EC$ Graphs

As remarked in Sect. 11.2, k_tEC graphs have yet to be characterized for any fixed $k \geq 3$. Sharp lower and upper bounds on the diameter of k_tEC graphs have however been obtained for $k = 3$ and $k = 4$. The following lemmas improve on the bound $\text{diam}(G) \leq 2\gamma_t(G) - 1$, which follows from Theorem 2.17.

Lemma 11.8 ([203]). *If G is a 3_tEC graph, then* $2 \leq \text{diam}(G) \leq 3$.

Lemma 11.9 ([204]). *If G is a 4_tEC graph, then* $2 \leq \text{diam}(G) \leq 4$.

Sharp upper bounds on the diameter of γ_tEC graphs for small k can be found in [127].

Theorem 11.4 ([127]). *For $k \in \{3,4,5,6\}$, the maximum diameter of a k_tEC graph is the value given by the following table:*

k	3	4	5	6
diam	3	4	6	7

That is, for $k \in \{3,4,5,6\}$, the maximum diameter of a k_tEC graph is $\lfloor 3(k-1)/2 \rfloor$.

A trivial general upper bound on the maximum diameter of a γ_tEC graph can also be found in [127].

Theorem 11.5 ([127]). *If G is a k_tEC graph, then* $\text{diam}(G) \leq 2k-3$.

However it remains an open problem to determine the maximum diameter of a k_tEC graph. Using the general constructions presented in Sect. 11.2.2, the following result provides a lower bound on the maximum diameter of a k_tEC graph.

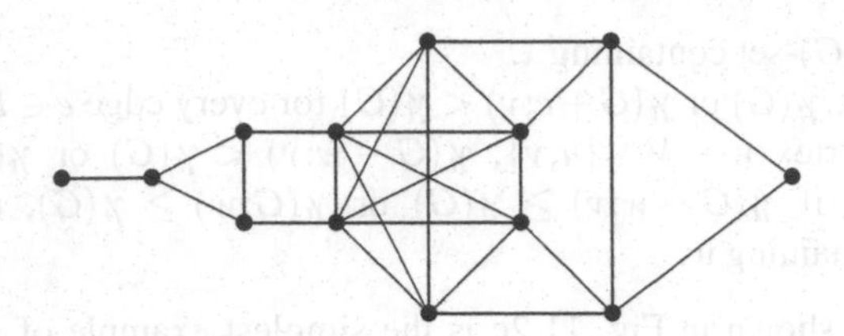

Fig. 11.5 The 5_tEC graph $G_3 \boxdot G_2$

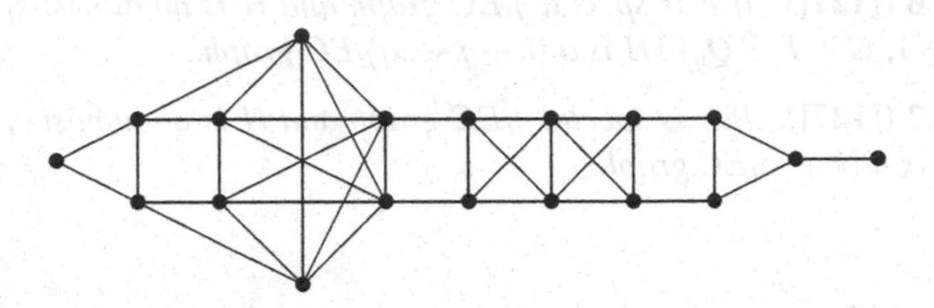

Fig. 11.6 The 7_tEC graph $G_2 \boxdot Q_1 \boxdot G_3$ with diameter 9

Theorem 11.6 ([127]). *For every $k \geq 2$, there exists a k_tEC graph with diameter at least $\lfloor 3(k-1)/2 \rfloor$.*

Proof. For $k \in \{3,4,5,6\}$, the desired result follows from Theorem 11.4. For completeness we will give examples for $k = 5$ and $k = 6$, even though the existence of these examples followed from Theorem 11.4. The graph $G_3 \boxdot G_2$ shown in Fig. 11.5 is a 5_tEC graph with diameter 6, thereby establishing the result for $k = 5$.

By Lemma 11.5, the graph $G_2 \boxdot G_2$ shown in Fig. 11.3 is a 6_tEC graph. Furthermore, $G_2 \boxdot G_2$ has diameter 7, thereby establishing the result for $k = 6$.

For $q \geq 1$, let $F_q = G_2 \boxdot Q_q \boxdot G_3$. As remarked earlier, G_2 is a special 3_tEC-graph while G_3 is an almost 2_tEC-graph. Hence by Lemma 11.6, the graph F_q is a k_tEC graph where $k = 5 + 2q \geq 7$. Furthermore, $\text{diam}(F_q) = 6 + 3q = 3(k-1)/2$. Hence for every $k \geq 7$ odd, there exists a k_tEC graph of diameter $3(k-1)/2$. (The 7_tEC graph $G_2 \boxdot Q_1 \boxdot G_3$ with diameter 9 is shown in Fig. 11.6.) This establishes the result for $k \geq 7$ odd.

For $q \geq 1$, let $H_q = G_2 \boxdot Q_q \boxdot G_2$. As remarked earlier, G_2 is a 3_tEC graph. Hence by Lemma 11.4, the graph H_q is a k_tEC graph where $k = 6 + 2q \geq 8$. Furthermore, $\text{diam}(H_q) = 7 + 3q = (3k-4)/2 = \lfloor 3(k-1)/2 \rfloor$. Hence for every $k \geq 8$ even, there exists a k_tEC graph of diameter $\lfloor 3(k-1)/2 \rfloor$. (The 8_tEC graph $G_2 \boxdot Q_1 \boxdot G_2$ of diameter 10 is shown in Fig. 11.4.) This establishes the result for $k \geq 8$ even and completes the proof of the theorem. □

11.2.4 Applications to Diameter-2-Critical Graphs

In this section, we discuss an important association with 3_tEC graphs and diameter-2-critical graphs. A graph G is called *diameter-2-critical* if its diameter is two, and the deletion of any edge increases the diameter. Diameter-2-critical graphs are extensively studied in the literature. Plesník [172] observed that all known diameter-2-critical graphs on n vertices have no more than $n^2/4$ edges and that the extremal graphs appear to be the complete bipartite graphs with partite sets whose cardinality differs by at most one. Murty and Simon (see [22]) independently made the following conjecture:

Conjecture 11.7 (Murty–Simon Conjecture). If G is a diameter-2-critical graph with n vertices and m edges, then $m \leq \lfloor n^2/4 \rfloor$, with equality if and only if G is the complete bipartite graph $K_{\lfloor \frac{n}{2} \rfloor, \lceil \frac{n}{2} \rceil}$.

According to Füredi [70], Erdös said that this conjecture goes back to the work of Ore in the 1960s. Plesník [172] proved that $m < 3n(n-1)/8$. Caccetta and Häggkvist [22] showed $m < .27n^2$. Fan [61] proved the first part of the conjecture for $n \leq 24$ and for $n = 26$. For $n \geq 25$, he obtained $m < n^2/4 + (n^2 - 16.2n + 56)/320 < .2532n^2$. Then Xu [206] gave an incorrect proof of the conjecture in 1984. In 1992 Füredi [70] gave an asymptotic result proving the conjecture is true for large n, that is, for $n > n_0$ where n_0 is a tower of 2's of height about 10^{14}. Although considerable work has been done in an attempt to completely resolve the conjecture and several impressive partial results have been obtained, the conjecture remains open for general n.

Conventionally the diameter of a disconnected graph is considered to be either undefined or defined as infinity. Here we use the former and hence require that a diameter-2-critical graph have minimum degree of two (since removing an edge incident to a vertex of degree one results in a disconnected graph). We note however that if we relaxed the condition and defined the diameter of disconnected graphs to be infinity, the only additional diameter-2-critical graphs are stars with order at least three.

Hanson and Wang [78] were the first to observe the following key relationship between diameter-2-critical graphs and total domination edge-critical graphs. Note that this relationship is contingent on total domination being defined in the complement $\overline{G}$ of the diameter-2-critical graph G, that is, $\overline{G}$ has no isolated vertices. Hence our elimination of stars as diameter-2-graphs is necessary for this association.

Theorem 11.8 ([78]). *A graph is diameter-2-critical if and only if its complement is 3_tEC or 4_t-supercritical.*

To illustrate Theorem 11.8, the graph H_5 in Fig. 11.7a is diameter-2-critical, and so by Theorems 11.8 and 11.2, its complement, $\overline{H_5}$, is 3_tEC

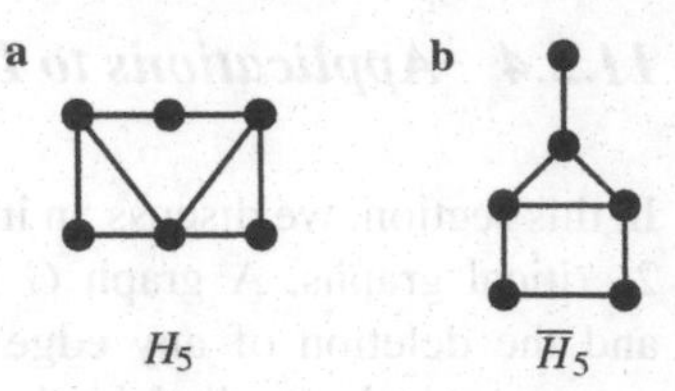

Fig. 11.7 The graph H_5 and its complement $\overline{H}_5$

By Theorem 11.2 in Sect. 11.2.1, a graph G is 4_t-supercritical if and only if G is the disjoint union of two nontrivial complete graphs. A simple calculus argument shows that the size of such a graph G is minimized when the two complete components of G have orders $\lfloor \frac{n}{2} \rfloor$ and $\lceil \frac{n}{2} \rceil$, respectively. Hence we have the following immediate corollary of Theorem 11.2.

Corollary 11.9 ([202]). *If G is a 4_t-supercritical graph of order n of minimum size, then $G = K_{\lfloor \frac{n}{2} \rfloor} \cup K_{\lceil \frac{n}{2} \rceil}$.*

By Corollary 11.9, Conjecture 11.7 holds for the case when G is a 4_t-supercritical graph and a subset of the complements of 4_t-supercritical graphs yield the extremal graphs of the conjecture. As the complement of a complete bipartite graph is not a 3_tEC graph, Theorems 11.8 and 11.2 imply that Conjecture 11.7 is equivalent to the following conjecture.

Conjecture 11.10. If G is a 3_tEC graph with order n and size m, then $m > \binom{n}{2} - \lfloor n^2/4 \rfloor = \lceil n(n-2)/4 \rceil$.

Recall that by Lemma 11.8, if G is a 3_tEC graph, then $2 \le \text{diam}(G) \le 3$. Hanson and Wang [78] proved the following result.

Theorem 11.11 ([78]). *If G is a 3_tEC graph of diameter three and of order n and size m, then $m \ge n(n-2)/4$.*

In order to prove that Conjecture 11.10 holds for 3_tEC graphs of diameter three, we need strict inequality in Theorem 11.11. Hence an additional edge is necessary to prove the conjecture in this case. Without strict inequality in Theorem 11.11, the conjecture is not proven. Indeed a surprising amount of work (see [94]) is required to find this one additional edge in the case when n is even. A slight adaption of the proof in [94] handles the case when n is odd as shown in [97]. We remark that the odd order case is proven independently by Wang (personal communication).

We summarize the graph classes, for which Conjecture 11.10 is known to hold in Table 11.1, and discuss these results below.

Connectivity-1 Graphs. Suppose G is a 3_tEC graph of order n and size m that contains a cut-vertex v. If $\text{diam}(G) = 2$, then the vertex v dominates $V(G)$, and so $\{v, v'\}$ is a TD-set in G where v' is any neighbor of v, contradicting the fact that $\gamma_t(G) = 3$. Hence, $\text{diam}(G) = 3$ and the desired result follows from the diameter-3 result in [94].

Connectivity-2 and Connectivity-3 Graphs. We remark that Conjecture 11.10 was proven true for graphs of connectivity-2 and connectivity-3 for even order n

Table 11.1 Graph classes for which Conjecture 11.10 holds

Conjecture 11.10 holds for the following graph classes	
Diameter-3 graphs	[94, 97] (Wang, P.[a])
Claw-free graphs	[98]
Connectivity-1 graphs	[94]
Connectivity-2 graphs	[97, 99] (Wang, P.[b])
Connectivity-3 graphs	[97, 99] (Wang, P.[c])
Bull-free graphs	[91]
Diamond-free graphs	[90]
C_4-free graphs	[89]
House-free graphs	[91]

[a]personal communication.
[b]personal communication.
[c]personal communication.

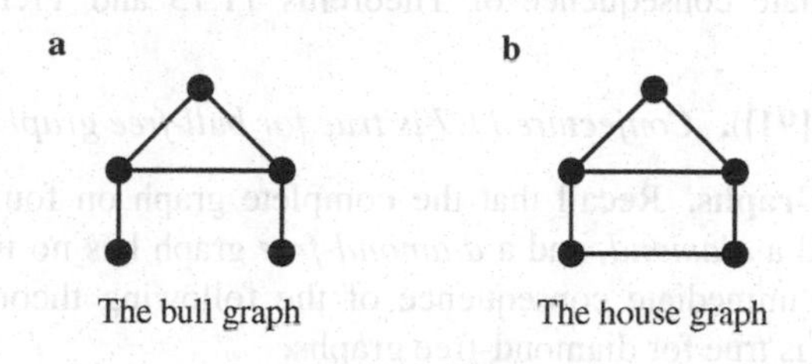

Fig. 11.8 The bull graph and the house graph

in [99]. A slight adaption of the proof in [94] handles the case when n is odd as shown in [97]. We remark that the odd order case is proven independently by Wang (personal communication).

Bull-Free Graphs. A *bull graph* consists of a triangle with two disjoint pendant edges as illustrated in Fig. 11.8a. A graph is *bull-free* if it has no bull as an induced subgraph. We observe that the bull is a self-complementary graph.

An independent set of size three is called a *triad*. A graph is *triad-free* if it contains no triad. The following result is proven in [91].

Theorem 11.12 ([91]). *Let G be a $3_t EC$ graph or a 4_t-supercritical graph. Then, G is bull-free if and only if G is triad-free.*

Recall that a clique of size three is called a *triangle* and a graph is *triangle-free* if it contains no triangle. Hence a graph G is *triad-free* if and only if its complement $\overline{G}$ is *triangle-free*. As an immediate consequence of Theorems 11.8 and 11.12 we therefore have the following result.

Theorem 11.13 ([91]). *Let G be a diameter-2-critical graph. Then, G is bull-free if and only if G is triangle-free.*

As a special case of a classic result of Turán [199] in 1941, we have the following bound on the maximum number of edges in a triangle-free graph.

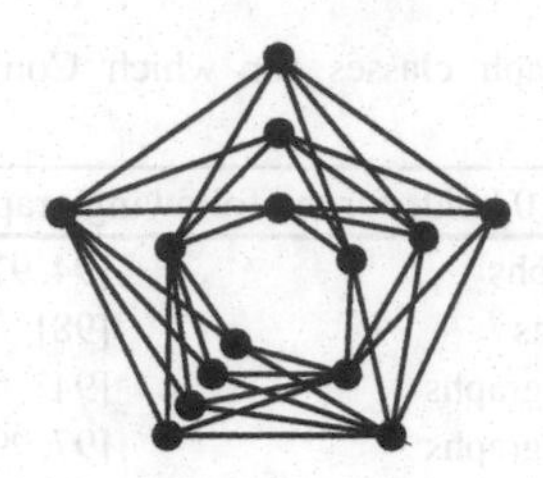

Fig. 11.9 A graph in the family $\mathscr{F}'$

Theorem 11.14 (Turán's Theorem). *If G is a triangle-free graph with n vertices and m edges, then $m \leq \lfloor n^2/4 \rfloor$, with equality if and only if G is the complete bipartite graph $K_{\lfloor \frac{n}{2} \rfloor, \lceil \frac{n}{2} \rceil}$.*

As an immediate consequence of Theorems 11.13 and 11.14, we have the following result.

Theorem 11.15 ([91]). *Conjecture 11.7 is true for bull-free graphs.*

Diamond-Free Graphs. Recall that the complete graph on four vertices minus one edge is called a *diamond*, and a *diamond-free* graph has no induced diamond subgraph. As an immediate consequence of the following theorem we see that Conjecture 11.10 is true for diamond-free graphs.

Theorem 11.16 ([90]). *Let G be a diamond-free graph. Then, G is a 3_tEC graph if and only if G is the cycle C_5 or the graph $\overline{H}_5$ in Fig. 11.7b.*

C_4-free Graphs. Recall that a graph is C_4-free if it does not contain C_4 as an induced subgraph. Let $\mathscr{G}_{C_4\text{-Free}}$ denote the family of graphs that can be obtained from a 5-cycle $v_1v_2v_3v_4v_5v_1$ by replacing each vertex v_i, $1 \leq i \leq 5$, with a clique A_i and adding all edges between A_i and A_{i+1}, where addition is taken modulo 5. As an immediate consequence of the following theorem we see that Conjecture 11.10 is true for C_4-free graphs.

Theorem 11.17 ([89]). *Let G be a C_4-free graph. Then, G is a 3_tEC graph if and only if $G \in \mathscr{G}_{C_4\text{-Free}}$.*

We will now relate Theorem 11.17 to the Murty–Simon Conjecture.

We remark that if a graph is C_k-free for some $k \geq 4$, then we say that it has no *hole* of length k and equivalently its complement $\overline{G}$ has no *antihole* of length k. As a consequence of Theorem 11.17, we have a characterization of diameter-2-critical graphs with no antihole of length 4. To state this characterization, we let $\mathscr{F}'$ denote the family of graphs that can be obtained from a 5-cycle $v_1v_2v_3v_4v_5v_1$ by replacing each vertex v_i, $1 \leq i \leq 5$, with a nonempty independent set A_i and adding all edges between A_i and A_{i+1}, where addition is taken modulo 5. Thus, $G \in \mathscr{G}_{C_4\text{-Free}}$ if and only if $\overline{G} \in \mathscr{F}'$. A graph in the family $\mathscr{F}'$ is illustrated in Fig. 11.9.

Corollary 11.18 ([90]). *Let G be a graph with no antihole of length 4. Then, G is a diameter-2-critical graph if and only if G is a complete bipartite graph with minimum degree at least 2 or $G \in \mathscr{F}'$.*

Since every graph in the family $\mathscr{F}'$ is triangle-free, as an immediate consequence of Theorems 11.18 and 11.14, we have the following result.

Corollary 11.19 ([90]). *Conjecture 11.7 holds for graphs with no antihole of length 4.*

House-Free Graphs. The complement of the path P_5 is the *house graph* as illustrated in Fig. 11.8b. A graph is *house-free* if it has no house as an induced subgraph or, equivalently, if its complement is P_5-free. The following result is proven in [92].

Theorem 11.20 ([92]). *Let G be a 3_tEC graph. Then, G is house-free if and only if $G \in \mathscr{G}_{C_4\text{-Free}}$.*

As an immediate consequence of Theorems 11.8 and 11.20, we therefore have the following result.

Theorem 11.21 ([92]). *Let G be P_5-free graph. Then, G is a diameter-2-critical graph if and only if G is a complete bipartite graph with minimum degree at least* 2 *or $G \in \mathscr{F}'$.*

As observed earlier, $G \in \mathscr{G}_{C_4\text{-Free}}$ if and only if $\overline{G} \in \mathscr{F}'$. Further, every graph in the family $\mathscr{F}'$ is triangle-free. Hence as a consequence of Theorems 11.14 and 11.21, we have the following result.

Corollary 11.22 ([92]). *Conjecture 11.7 is true for P_5-free graphs.*

11.2.4.1 Properties of 3_tEC Graphs

It is shown in [95] that every 3_tEC graph, except for the 5-cycle, contains a triangle. As a consequence, no bipartite graph is 3_tEC. The tripartite 3_tEC graphs are characterized in [95], and it is shown that there are exactly five such graphs (one each of orders 5, 8, and 9 and two of order 6).

Define $f(r)$ as follows. By a 3-path in a tournament, we mean an oriented (or directed) path on three vertices (with two arcs). Let $f(r)$ denote the smallest integer such that all r-arc-colored tournaments of order at least $f(r)$ contain a monochromatic 3-path. Using Ramsey theory we note that $f(r)$ exists as any monochromatic K_3 in the underlying graph of a tournament (which is a complete graph) contains a 3-path. It is shown in [95] that there are only a finite number of 3_tEC k-partite graphs.

Theorem 11.23 ([95]). *There are no 3_tEC k-partite graphs with a partite set of size $f(k-1)$ or larger for any $k \geq 2$.*

Corollary 11.24 ([95]). *A 3_tEC k-partite graph has order at most $k(f(k-1)-1)$.*

For every $k \geq 3$, a 3_tEC k-partite graph G of order $2(k+1)$ is constructed in [95] as follows. Let $V(G) = A \cup B \cup C$, where $A = \{a_B, a_C, a\}$, $B = \{b_A, b_C, b\}$, $C_i = \{c_i^1, c_i^2\}$ for $i = 1, \ldots, k-2$ and $C = \cup_{i=1}^{k-2} C_i$. For $j = 1, 2$, let $C^j = \cup_{i=1}^{k-2}\{c_i^j\}$. Add edges so that the set C^1 induces a clique and the set C^2 induces a clique. Also add edges so that $a_B \succ B \cup C^2$, $a_C \succ C$, $a \succ C^1$, and so that $b_A \succ A \cup C^2$, $b_C \succ C$, and $b \succ C^1$. Finally, add the edges ab_C and ba_C. Then as shown in [95], G is 3_tEC k-partite graph of order $2(k+1)$.

As remarked earlier, the 5-cycle is the only 3_tEC K_3-free graph. It is also shown in [95] that there are only a finite number of 3_tEC K_4-free graphs and they conjecture that there are only a finite number of 3_tEC K_p-free graphs for each $p \geq 3$.

As remarked earlier, the problem to characterize 3_tEC graphs remains open. Hence it is important to study properties of these graphs. Van der Merwe et al. [203] characterized the 3_tEC graphs with end-vertices.

Theorem 11.25 ([203]). *Let G be a graph with an end-vertex u that is adjacent to a vertex v. Let $A = N(v) \setminus \{u\}$ and let $B = V \setminus N[v]$. Then, G is 3_tEC if and only if A is a clique and $|A| \geq 2$, B is a clique and $|B| \geq 2$, and every vertex in A is adjacent to $|B|-1$ vertices in B and every vertex in B is adjacent to at least one vertex in A.*

Theorem 11.25 implies that Conjecture 11.10 holds when $\delta(G) = 1$. In fact, Conjecture 11.10 holds when $\delta(G) \leq 0.3n$ as the following result shows.

Theorem 11.26 ([96]). *Let G be a 3_t-critical graph of order n and size m. Let $\delta = \delta(G) \geq 1$. Then the following hold:*

(a) *If $\delta \leq 0.3n$, then $m > n(n-2)/4$.*
(b) *If $n \geq 2000$ and $\delta \leq 0.321n$, then $m > n(n-2)/4$.*

Since $\delta(G) = n - 1 - \Delta(\overline{G})$, as a consequence of Theorem 11.26 we have that Conjecture 11.7 holds when $\Delta(G) \geq 0.7n$.

Theorem 11.27 ([96]). *Let G be a diameter-2-critical of order n and size m. Let $\Delta = \Delta(G)$. Then the following hold:*

(a) *If $\Delta \geq 0.7n$, then $m < \lfloor n^2/4 \rfloor$.*
(b) *If $n \geq 2000$ and $\Delta \geq 0.6787n$, then $m < \lfloor n^2/4 \rfloor$.*

Theorem 11.28 ([203]). *Every 3_tEC graph with no end-vertex is 2-connected.*

Matching properties in 3_tEC graphs are studied in [136]. The authors in [136] prove the following result.

Theorem 11.29 ([136]). *Every 3_tEC 2-connected graph of even order has a perfect matching, while every 3_tEC 2-connected graph of odd order is factor-critical.*

11.2.5 Total Domination Vertex-Critical Graphs

A vertex v in a graph G is *γ_t-vertex-critical* if $\gamma_t(G-v) < \gamma_t(G)$. Since total domination is undefined for a graph with isolated vertices, we say that a graph G is *total domination vertex-critical*, abbreviated γ_tVC, if every vertex of G that is not adjacent to a vertex of degree one is γ_t-vertex-critical. In particular, if $\delta(G) \geq 2$, then G is γ_tVC if every vertex of G is γ_t-vertex-critical. Further if G is γ_tVC and $\gamma_t(G) = k$, we say that G is *k_tVC*. Thus if G is a k_tVC graph, then its total domination number is k and the removal of any vertex not adjacent to a vertex of degree one decreases the total domination number. For example, the 5-cycle is 3_tVC as is the complement of the Petersen graph. The γ_tVC graphs with end-vertices are characterized in [74].

Theorem 11.30 ([74]). *Let G be a connected graph of order at least 3 with at least one end-vertex. Then, G is k_tVC if and only if G is the corona,* $\text{cor}(H)$, *of some connected graph H of order k with $\delta(H) \geq 2$.*

A graph H is *vertex diameter k-critical* if $\text{diam}(H) = k$ and $\text{diam}(H-v) > k$ for all $v \in V(H)$. Hanson and Wang [78] observed the following result.

Theorem 11.31 ([78]). *For a graph G, $\gamma_t(G) = 2$ if and only* $\text{diam}(\overline{G}) > 2$.

The following characterization of 3_tVC graphs is given in [74].

Theorem 11.32 ([74]). *A connected graph G is 3_tVC if and only if $\overline{G}$ is vertex diameter-2-critical or G is the net,* $\text{cor}(K_3)$.

For example, the Petersen graph is vertex diameter-2-critical, and so the complement is 3_tVC. Bounds on the diameter of a connected k_tVC graph are established in [74].

Theorem 11.33 ([74]). *Let G_k be a connected k_tVC graph of maximum diameter. For $k \geq 9$, $\text{diam}(G_k) \leq 2k-3$. For $k \leq 8$, the diameter of G_k is the value given by the following table.*

k	3	4	5	6	7	8
diam	3	4	6	7	9	11

Theorem 11.34 ([74]). *For all $k \equiv 2 \pmod 3$, there exists a k_tVC graph of diameter* $(5k-7)/3$.

It remains, however, an open problem to determine the maximum diameter of a k_tVC graph. It is possible that for sufficiently large k, the maximum diameter is $(5k-7)/3$.

Wang et al. [205] studied matching properties in total domination vertex-critical graphs. They proved the following result.

Theorem 11.35 ([205]). *Let G be a 3_tVC graph that is $K_{1,5}$-free. If G has even order, then G has a perfect matching. If G has odd order, then G is factor-critical.*

Wang et al. [205] pose the following question. Can the hypothesis that G is $K_{1,5}$-free in Theorem 11.35 be lowered? This question is answered in the affirmative in [136] where it is shown that the $K_{1,5}$-free condition in Theorem 11.35 can be relaxed to $K_{1,7}$-free for even order graphs and relaxed to $K_{1,6}$-free for odd order graphs.

Theorem 11.36 ([136]). *Let G be a 3_tVC graph with minimum degree at least two. Then the following hold:*

(a) *If G has odd order and is $K_{1,6}$-free, then G is factor-critical.*
(b) *If G has even order and is $K_{1,7}$-free, then G has a perfect matching.*

To show that the results in Theorem 11.36 are best possible, it is shown in [136] that there exists a 3_tVC graph of odd order with minimum degree at least five that is $K_{1,7}$-free and is not factor-critical. Further it is shown in [136] that there exists a 3_tVC graph of even order with minimum degree at least five and with no perfect matching that is $K_{1,8}$-free.

11.2.6 γ_tVC Graphs of High Connectivity

We remark that our terminology used for total domination vertex-critical graphs is similar to that used for domination vertex-critical graphs. For example, a graph G is *k-γ-critical* if $\gamma(G) = k$ and $\gamma(G - v) < k$ for all vertices v in G. A graph is *γ-vertex-critical*, abbreviated γVC, if it is k-γ-critical for some k. The study of γVC graphs was begun by Brigham, Chinn, and Dutton [21] and is now well studied in the literature.

In [74], the authors list as an open problem the connection between γVC and γ_tVC graphs. For example, the Cartesian product $K_3 \Box K_3$ is γVC but not γ_tVC. The cycle C_5 is γ_tVC, but not γVC. For $k \geq 1$, the cycle C_{12k+1} is both γVC and γ_tVC. The authors in [74] pose the following question. Which graphs are γVC and γ_tVC or one but not the other? In order to provide a partial answer to this question, the authors in [122] consider the family of Harary graphs.

For $2 \leq k < n$, the Harary graph $H_{k,n}$ on n vertices is defined as follows. Place n vertices around a circle, equally spaced. If k is even, $H_{k,n}$ is formed by making each vertex adjacent to the nearest $k/2$ vertices in each direction around the circle. If k is odd and n is even, $H_{k,n}$ is formed by making each vertex adjacent to the nearest $(k-1)/2$ vertices in each direction around the circle and to the diametrically opposite vertex. In both cases, $H_{k,n}$ is k-regular. If both k and n are odd, $H_{k,n}$ is constructed as follows. It has vertices $0, 1, \ldots, n-1$ and is constructed from $H_{k-1,n}$ by adding edges joining vertex i to vertex $i + (n-1)/2$ for $0 \leq i \leq (n-1)/2$. Harary [79] showed that for $2 \leq k < n$, the graph $H_{k,n}$ is k-connected.

Total dominating sets in Harary graphs are studied by Khodkar, Mojdeh, and Kazemi [152]. The following class of γ_tVC graphs of high connectivity is given in [122].

Theorem 11.37 ([122]). *For $\ell \geq 1$ and $k \geq 2$, the Harary graph $H_{2k+1,2\ell(2k+1)+2}$ is a $(2k+1)$-connected graph that is $(2\ell+2)_t VC$.*

We remark that as a special case of Theorem 11.37 when $\ell = 1$, the Harary graph $H_{2k+1,4k+4}$ is a $(2k+1)$-connected graph that is a $4_t VC$ graph of diameter 2 for every integer $k \geq 2$. A different family of $4_t VC$ graphs of diameter 2 is constructed by Loizeaux and van der Merwe [163].

Theorem 11.38 ([122]). *For $\ell \geq 1$ and $k \geq 2$, the Harary graph $H_{2k,\ell(3k+1)+1}$ is a $2k$-connected graph of diameter 2 that is $(2\ell+1)_t VC$.*

Using results from [122] and [152], we have the following partial answer to the question posed in [74] asking which graphs are γVC and $\gamma_t VC$ or one but not the other.

Theorem 11.39. *For integers $k \geq \ell - 2 \geq 0$ where $k - \ell + 2 = r$ and $r \in \{0,1\}$ or $k+2 \leq r \leq 2k+1$, the Harary graph $H_{2k+1,2\ell(2k+1)+2}$ is $\gamma_t VC$ and is not γVC.*

Theorem 11.40. *For all integers $\ell \geq 1$ and $k \geq 2$ such that $\ell(3k+1)$ is not divisible by $2k+1$, the Harary graph $H_{2k+1,2\ell(2k+1)+2}$ is $\gamma_t VC$ and is not γVC.*

11.3 Total Domination Edge Addition Stable Graphs

Adding an edge to a graph cannot increase the total domination number. Hence, $\gamma_t(G+e) \leq \gamma_t(G)$ for every edge $e \in E(\overline{G}) \neq \emptyset$. A graph G is said to be *total domination edge addition stable*, or γ_t^+*-stable* for short, if $\gamma_t(G+e) = \gamma_t(G)$ for every edge $e \in E(\overline{G}) \neq \emptyset$. Further if $\gamma_t(G) = k$, then we say that G is a k_t^+-stable graph. Thus, if G is k_t^+-stable, then its total domination number is k and the addition of any edge leaves the total domination number unchanged. We remark that by definition complete graphs are not γ_t^+-stable.

Every graph G with $\gamma_t(G) = 2$ that is not a complete graph is 2_t^+-stable. Hence, we are only interested in k_t^+-stable graphs where $k \geq 3$. The Petersen Graph is an example of a 4_t^+-stable graph. For a more general family of γ_t^+-stable graphs, it is shown in [84] that for every integer $k \geq 1$, the path P_{4k} is $(2k)_t^+$-stable. We also remark that for every integer $k \geq 1$, the cycle C_{4k} is $(2k)_t^+$-stable. More generally, we have the following result.

Proposition 11.41 ([45]). *For every integer $n \geq 8$ and every even integer k such that $4 \leq k \leq n/2$, there exists a connected γ_t^+-stable graph G of order n such that $\gamma_t(G) = k$.*

Proposition 11.42 ([45]). *For every odd integer $k \geq 3$ and any integer $n \geq 2k+2$, there exists a connected γ_t^+-stable graph G with $\gamma_t(G) = k$.*

To illustrate Proposition 11.42, recall that a *caterpillar* is a tree such that the removal of all its leaves produces a path, called its *spine*. The *code* of the caterpillar having spine $P_s = (v_1, v_2, \cdots, v_s)$ is the ordered s-tuple $(\ell_1, \ell_2, \cdots, \ell_s)$, where ℓ_i is the number of leaves adjacent to v_i. Let $k \geq 3$ be an odd integer, and let $n \geq 2k+2$, and so $3 \leq k \leq (n-2)/2$. To construct a connected γ_t^+-stable graph G with $\gamma_t(G) = k$, we begin with a caterpillar G_1 having spine (y,x,z) and code $(2,1,n-2k)$. If $k=3$, then we let $G = G_1$. If $k \geq 5$, then let G be obtained from the above caterpillar, G_1, by adding a path P_{2k-6} and adding an edge from a leaf of this path to a leaf of the vertex y in G_1. The resulting graph G is a connected γ_t^+-stable graph G with $\gamma_t(G) = k$ as shown in [45].

γ_t^+-Stable graphs having a specified total domination number and induced subgraph are constructed in [45].

Proposition 11.43 ([45]). *For any integer $k \geq 2$ and any graph H, there exists a γ_t^+-stable graph G such that $\gamma_t(G) = k$ and H is a vertex-induced subgraph of G.*

As a consequence of Proposition 11.43 we remark that there exists no forbidden subgraph characterization of γ_t^+-stable graphs. A characterization of 3_t^+-stable graphs is given in [45].

Theorem 11.44 ([45]). *A graph G is 3_t^+-stable if and only if $\gamma(G) = 3$ and for any $\gamma_t(G)$-set S and for every $v \in S$, $|\text{epn}(v,S)| \geq 1$ and $|\text{pn}(v,S)| \geq 2$.*

A characterization of k_t^+-stable graphs for $k \geq 4$ is also given in [45]. To state this result, for a graph G and for any two vertices $u,v \in V(G)$ such that $\{u,v\}$ does not dominate G, we define G_{uv} to be the graph

$$G_{uv} = G[V(G) \setminus (N[u] \cup N[v])].$$

Theorem 11.45 ([45]). *Let $k \geq 4$ be an integer. A graph G is k^+-stable if and only if the following two conditions hold.*

1. *For any $\gamma_t(G)$-set S and for every $v \in S$, one of the following is true.*
   (a) $|\text{epn}(v,S)| \geq 1$ *and* $|\text{pn}(v,S)| \geq 2$.
   (b) $\text{epn}(v,S) = \emptyset$ *and* $|\text{ipn}(v,S)| \geq 3$.
2. *For every two vertices $u,v \in V(G)$, such that $d_G(u,v) \geq 3$ and G_{uv} is isolate-free, we have $\gamma_t(G_{uv}) > \gamma_t(G) - 3$.*

We remark that if G is a graph with $\text{diam}(G) = 2$, then Condition 2 in Theorem 11.45 is vacuously satisfied. Additionally, Condition 2 is also vacuously satisfied if $\gamma_t(G) = 4$. Hence as an immediate consequence of Theorem 11.45 we have the following result.

Corollary 11.46 ([45]). *Let G be a graph with $\gamma_t(G) = 4$ or with* $\text{diam}(G) = 2$ *and $\gamma_t(G) \geq 5$. The graph G is γ_t^+-stable if and only if for every $\gamma_t(G)$-set S and for every $v \in S$, one of the following properties hold:*

(a) $|\text{epn}(v,S)| \geq 1$ *and* $|\text{pn}(v,S)| \geq 2$.
(b) $\text{epn}(v,S) = \emptyset$ *and* $|\text{ipn}(v,S)| \geq 3$.

Upper bounds on the total domination number of γ_t^+-stable graphs are determined in [45]. If we add the condition that a graph G is γ_t^+-stable, then the bound in Theorems 5.5 and 5.7 can be improved.

Theorem 11.47 ([45]). *Let G be a connected γ_t^+-stable graph of order n with $\gamma_t(G) \geq 3$. Then, $\gamma_t(G) \leq n/2$. Further if $\delta(G) \geq 2$, then equality holds if and only if $G = C_n$ where $n \equiv 0 \pmod 4$, implying that if $\delta(G) \geq 3$, then $\gamma_t(G) < n/2$.*

Theorem 11.48 ([45]). *Let G be a connected γ_t^+-stable graph of order n with $\gamma_t(G) \geq 3$ and with maximum degree Δ. Then the following hold:*

(a) *If $\delta \geq 3$, then $\gamma_t(G) \leq \left(\frac{\Delta}{2\Delta+1}\right) n$.*
(b) *If $\delta \geq 4$, then $\gamma_t(G) \leq \left(\frac{\Delta}{2\Delta+2}\right) n$.*

As an immediate consequence of Theorem 11.48, we have the following result.

Corollary 11.49 ([45]). *Let G be a connected γ_t^+-stable graph of order n with $\gamma_t(G) \geq 3$. Then the following hold:*

(a) *If G is a cubic graph, then $\gamma_t(G) \leq 3n/7$.*
(b) *If G is a 4-regular graph, then $\gamma_t(G) \leq 2n/5$.*

Upper and lower bounds for the total domination number of γ_t^+-stable claw-free cubic graphs are determined in [45].

Theorem 11.50 ([45]). *If G is a γ_t^+-stable claw-free cubic graph of order n with $\gamma_t(G) \geq 3$, then $n/3 \leq \gamma_t(G) \leq 2n/5$.*

11.4 Total Domination Edge Removal Stable Graphs

A graph G is said to be *total domination edge removal stable*, or γ_t^-*-stable* for short, if $\gamma_t(G-e) = \gamma_t(G)$ for every edge $e \in E(G)$. Further if $\gamma_t(G) = k$, then we say that G is a k_t^--stable graph. Thus if G is k_t^--stable, then its total domination number is k and the removal of any edge leave the total domination number unchanged. Since a path and a cycle on $n \geq 3$ vertices have equal total domination numbers, every cycle is a γ_t^--stable graph. By convention, if G is a graph with an isolated vertex, then we define $\gamma_t(G) = \infty$. Thus if e is an edge of a graph G incident with a leaf, then $\gamma_t(G-e) = \infty$.

The following result is established in [46].

Proposition 11.51 ([46]). *A graph G is γ_t^--stable if and only if $\delta(G) \geq 2$, and for each $e = uv \in E(G)$, there exists a $\gamma_t(G)$-set S such that one of the following conditions is satisfied:*

(a) *$u, v \notin S$.*
(b) *$u, v \in S$, $|N(u) \cap S| \geq 2$ and $|N(v) \cap S| \geq 2$.*
(c) *Without loss of generality, if $u \in S$ and $v \notin S$, then $|N(v) \cap S| \geq 2$.*

For a graph G, define $T_t(G) = \{v \in V(G) \mid v$ belongs to some $\gamma_t(G)$-set$\}$. The following result characterizes bipartite γ_t^--stable graphs.

Theorem 11.52 ([46]). *Let G be a bipartite graph. Then, G is γ_t^--stable if and only if for every vertex v in G, $|N(v) \cap T_t(G)| \geq 2$.*

As an immediate consequence of Theorem 11.52, we note that if a bipartite graph G has two disjoint $\gamma_t(G)$-sets, then G is γ_t^--stable.

The minimum number of edges required for a total domination stable graph in terms of its order and total domination number is determined in [47].

Theorem 11.53 ([47]). *If G is a connected γ_t^--stable graph of order n, size m, with total domination number γ_t, and with minimum degree $\delta \geq 2$, then*

$$m \geq \begin{cases} \frac{1}{2}((\delta+1)n - \delta\gamma_t) & \text{if } \gamma_t \text{ is even} \\ \frac{1}{2}((\delta+1)n - \delta\gamma_t + 1) & \text{if } \gamma_t \text{ is odd.} \end{cases}$$

Further for every given integer $\delta \geq 2$ and every given integer $\gamma_t \geq 2$, there exists a connected γ_t^--stable graph that is δ-regular of order n, size m, and with total domination number γ_t that achieves the above lower bound.

Theorem 11.54 ([47]). *Let G be a connected γ_t^--stable graph of order n, size m, and with total domination number γ_t. Then the following hold.*

(a) *If γ_t is even, then $m \geq 3n/2 - \gamma_t$ with equality if and only if $G = C_n$ and $n \equiv 0 \pmod 4$.*
(b) *If γ_t is odd, then $m \geq (3n+1)/2 - \gamma_t$ with equality if and only if $G = C_n$ and n is odd.*

11.5 Total Domination Vertex Removal Changing Graphs

We define a graph to be *total domination vertex removal changing*, or γ_t^-*-vertex changing* for short, if $\gamma_t(G - v) \neq \gamma_t(G)$ for any arbitrary $v \in V(G)$. For a graph G, we let $S(G)$ denote the set of support vertices of G. By convention, if G is a graph with an isolated vertex, then we define $\gamma_t(G) = \infty$. Thus if $v \in S(G)$, then $\gamma_t(G - v) = \infty$. Given a graph G, we define a partition on $V(G)$ such that $V(G) = V^0(G) \cup V^+(G) \cup V^-(G)$, where

- $V^0(G) = \{v \in V(G) \mid \gamma_t(G - v) = \gamma_t(G)\}$
- $V^+(G) = \{v \in V(G) \mid \gamma_t(G - v) > \gamma_t(G)\}$
- $V^-(G) = \{v \in V(G) \mid \gamma_t(G - v) < \gamma_t(G)\}$

Thus a graph G is γ_t^--vertex changing if $V(G) = V^+(G) \cup V^-(G)$. If G is a γ_t^--vertex changing graph with $V(G) = V^+(G)$, then it is shown in [48] that $G = mK_2$. Hence, we focus our attention on γ_t^--vertex changing graphs G with $V(G) \neq V^+(G)$.

Let $\mathscr{F}_{change}$ be the family of γ_t^--vertex changing graphs defined in [48] as follows. A graph $G \in \mathscr{F}_{change}$ if G can be obtained from a connected graph H, where every support vertex of H is a strong support vertex, by adding for every nonsupport vertex $v \in V(H)$, a new vertex v', and the edge vv'.

Theorem 11.55 ([48]). *Let G be a connected graph with $V(G) \neq V^+(G)$. Then, G is a γ_t^--vertex changing graph if and only if $G \in \mathscr{F}_{change}$.*

Bounds on the total domination number of a γ_t^--vertex changing graph in terms of its order are obtained in [48]. For this purpose, the authors in [48] define a family $\mathscr{H}'$ of graphs. Let $\mathscr{H}'$ be the family of all graphs G that can be obtained from a connected graph F, by adding to every vertex v in F two disjoint copies of K_2 and adding an edge from v to one vertex in each copy of K_2.

Theorem 11.56 ([48]). *If G is a connected γ_t^--vertex changing graph of order n with $V^+(G) \neq \emptyset$ and $V^-(G) \neq \emptyset$, then $n/2 \leq \gamma_t(G) \leq 3n/5$. Furthermore, the following hold:*

(a) *$\gamma_t(G) = n/2$ if and only if $G = H \square K_1$ for some connected graph H where $\delta(H) \geq 2$.*
(b) *$\gamma_t(G) = 3n/5$ if and only if $G \in \mathscr{H}'$.*

11.6 Total Domination Vertex Removal Stable Graphs

We define a graph to be *total domination vertex stable*, or *γ_t^--vertex stable* for short, if $\gamma_t(G - v) = \gamma_t(G)$ for every $v \in V(G)$. Thus using the notation introduced in Sect. 11.4, a graph G is γ_t^--vertex stable if $V(G) = V^0(G)$. The γ_t^--vertex stable graphs are characterized in [48]. Recall that in Sect. 4.4 the set $\mathscr{A}_t(G)$ of a graph G is defined as $\mathscr{A}_t(G) = \{v \in V(G) \mid v \text{ is in every } \gamma_t(G)\text{-set}\}$.

Theorem 11.57 ([48]). *A graph G is γ_t^--vertex stable if and only if $\delta(G) \geq 2$ and for every $v \in V(G)$, both of the following conditions hold:*

(a) *There is no $\gamma_t(G)$-set, S, such that $v \notin S$ and $\text{pn}(u, S) = \{v\}$ for some vertex $u \in S$.*
(b) *Either $v \notin A_t(G)$ or $v \in A_t(G)$ and there exist a TD-set S in $G - v$ such that $|S| = \gamma_t(G)$ and $S \subseteq V(G) \setminus N[v]$.*

Let $\mathcal{F}_{change}$ be the family of γ_t-vertex changing graphs defined in [48] as follows: A graph $G \in \mathcal{F}_{change}$ if G can be obtained from a connected graph H, where every support vertex of H is a strong support vertex, by adding for every nonsupport vertex $u \in V(H)$, a new vertex u', and the edge uu'.

Theorem 11.55 ([48]). *Let G be a connected graph with $V^{+}(G) \neq V^{-}(G)$. Then, G is a γ_t-vertex changing graph if and only if $G \in \mathcal{F}_{change}$.*

Bounds on the total domination number of a γ_t-vertex changing graph in terms of its order are obtained in [48]. For this purpose, the authors in [48] define a family $\mathcal{K}$ of graphs. Let $\mathcal{K}$ be the family of all graphs G that can be obtained from a connected graph F, by adding to every vertex v in F two disjoint copies of K_2 and adding an edge from v to one vertex in each copy of K_2.

Theorem 11.56 ([48]). *If G is a connected γ_t-vertex changing graph of order n with $V^{+}(G) \neq \emptyset$ and $V^{-}(G) \neq \emptyset$, then $n/2 \le \gamma_t(G) \le 3n/5$. Furthermore, the following hold:*

(a) $\gamma_t(G) = n/2$ if and only if $G = H \circ K_1$ for some connected graph H where $\delta(H) \ge 2$.
(b) $\gamma_t(G) = 3n/5$ if and only if $G \in \mathcal{K}$.

11.6 Total Domination Vertex Removal Stable Graphs

We define a graph to be *total domination vertex stable*, or γ_t-*vertex stable* for short, if $\gamma_t(G - v) = \gamma_t(G)$ for every $v \in V(G)$. Thus using the notation introduced in Sect. 11.4, a graph G is γ_t-vertex stable if $V(G) = V^0(G)$. The γ_t-vertex stable graphs are characterized in [48]. Recall that in Sect. 5.4 the set $\mathcal{A}_t(G)$ of a graph G is defined as $\mathcal{A}_t(G) = \{v \in V(G) \mid v$ is in every $\gamma_t(G)$-set$\}$.

Theorem 11.57 ([48]). *A graph G is γ_t-vertex stable if and only if $\delta(G) \ge 2$ and for every $v \in V(G)$, both of the following conditions hold:*

(a) There is no $\gamma_t(G)$-set S such that $v \notin S$ and $\text{pn}(u, S) = \{v\}$ for some vertex $u \in S$.
(b) Either $v \notin \mathcal{A}_t(G)$ or $v \in \mathcal{A}_t(G)$ and there exist a TD-set S in $G - v$ such that $|S| = \gamma_t(G)$ and $S \subseteq V(G) \setminus N[v]$.

Chapter 12
Total Domination and Graph Products

12.1 Introduction

The study of graphical invariants on graph products has resulted in several famous conjectures and open problems in graph theory. In this section, we investigate the behavior of the total domination number on the Cartesian product and the direct product of graphs. These products are both examples of graph products.

By a *graph product* $G \otimes H$ on graphs G and H, we mean the graph that has vertex set the Cartesian product of the vertex sets of G and H (i.e., $V(G \otimes H) = V(G) \times V(H)$) and edge set that is determined entirely by the adjacency relations of G and H. Exactly how it is determined depends on what kind of graph product we are considering. An outstanding survey of the behavior of the domination number on a graph product has been written by Nowakowski and Rall [169].

12.2 The Cartesian Product

Recall that for graphs G and H, the Cartesian product $G \,\square\, H$ is the graph with vertex set $V(G) \times V(H)$ where two vertices (u_1, v_1) and (u_2, v_2) are adjacent if and only if either $u_1 = u_2$ and $v_1 v_2 \in E(H)$ or $v_1 = v_2$ and $u_1 u_2 \in E(G)$. The most famous open problem involving domination in graphs is the more than four-decade-old conjecture of Vizing which states the domination number of the Cartesian product of any two graphs is at least as large as the product of their domination numbers. A survey of what is known about attacks on Vizing's conjecture can be found in [19]. Here we investigate a similar problem for total domination.

It is conjectured in [124] that the product of the total domination numbers of two graphs without isolated vertices is bounded above by twice the total domination number of their Cartesian product. This conjecture was solved by Pak Tung Ho [142].

M.A. Henning and A. Yeo, *Total Domination in Graphs*, Springer Monographs in Mathematics, DOI 10.1007/978-1-4614-6525-6_12,

Theorem 12.1 ([142]). *For graphs G and H without isolated vertices,*

$$\gamma_t(G)\gamma_t(H) \le 2\gamma_t(G \Box H).$$

Theorem 12.1 is perhaps a surprising result in the sense that the analogous problem for the ordinary domination number is arguably the most stubborn open problem in the area of domination theory. For this reason, we provide a proof of this important result.

The proof we provide of Theorem 12.1 uses what is called the *double-projection argument* by Clark and Suen [19,36] which nicely incorporates the product structure of the Cartesian product of two graphs.

First we need the notion of a fiber of a graph. For a vertex g of G, the subgraph of $G \Box H$ induced by the set $\{ (g,h) \mid h \in V(H) \}$ is called an *H-fiber* and is denoted by ${}^gH$. Similarly, for $h \in H$, the *G-fiber*, G^h, is the subgraph induced by $\{ (g,h) \mid g \in V(G) \}$. We note that all G-fibers are isomorphic to G and all H-fibers are isomorphic to H.

We will also have need of projection maps from the Cartesian product $G \Box H$ to one of the factors G or H. The *projection to H* is the map $p_H \colon V(G \Box H) \to V(H)$ defined by $p_H(g,h) = h$, while the *projection to G* is the map $p_G \colon V(G \Box H) \to V(G)$ defined by $p_G(g,h) = g$. We are now ready to present our proof of Theorem 12.1.

Proof. Let H be a graph with $\gamma_t(H) = k$ and let $\{h_1, h_2, \ldots, h_k\}$ be a $\gamma_t(H)$-set. Let $\{\pi_1, \pi_2, \ldots, \pi_k\}$ be a partition of $V(H)$ chosen so that $\pi_i \subseteq N_H(h_i)$ for each i, $1 \le i \le \gamma_t(H)$; that is, the set $\{h_i\}$ totally dominates the set π_i in H. For a vertex g of G we call the set of vertices $\{g\} \times \pi_i$ an *H-cell*. We illustrate an H-cell in Fig. 12.1.

Let $V = V(G \Box H)$ and let D be a $\gamma_t(G \Box H)$-set. Let $g \in V(G)$ and let $i \in \{1,2,\ldots,k\}$. If the set $D \cap V({}^gH)$ totally dominates all the vertices of the H-cell $\{g\} \times \pi_i$ in the H-fiber ${}^gH$, then we say that this H-cell is *totally dominated by D from within the corresponding H-fiber* and we call such a cell a *dominated H-cell*. We note that if at least one vertex in the H-cell $\{g\} \times \pi_i$ is not adjacent to a vertex of $D \cap V({}^gH)$ in the H-fiber ${}^gH$, then the H-cell $\{g\} \times \pi_i$ is not dominated by D from within the H-fiber ${}^gH$ and is therefore not a dominated H-cell. Further in this case, at least one vertex in the H-cell $\{g\} \times \pi_i$ is totally dominated by a vertex of $D \cap V(G_i)$ in $G \Box H$, implying that the vertex g is adjacent to a vertex of the projection $p_G(D \cap V(G_i))$ in the graph G. We now extend the projection set $p_G(D \cap V(G_i))$ to a TD-set in G as follows. For each vertex $g \in V(G)$ such that the H-cell $\{g\} \times \pi_i$ is a dominated H-cell, we add an arbitrary neighbor of g in G to the set $p_G(D \cap V(G_i))$. The resulting set is a TD-set in G of cardinality at most $|p_G(D \cap V(G_i))|$ plus the number of dominated H-cells $\{g\} \times \pi_i$. For $i = 1,2,\ldots,k$, let ℓ_i be the number of dominated H-cells in G_i. By the above arguments, we have that $|p_G(D \cap V(G_i))| + \ell_i \ge \gamma_t(G)$, implying that

$$|D| + \sum_{i=1}^{k} \ell_i \ge \sum_{i=1}^{k} (|p_G(D \cap V(G_i))| + \ell_i) \ge \sum_{i=1}^{k} \gamma_t(G) = \gamma_t(G)\gamma_t(H). \tag{12.1}$$

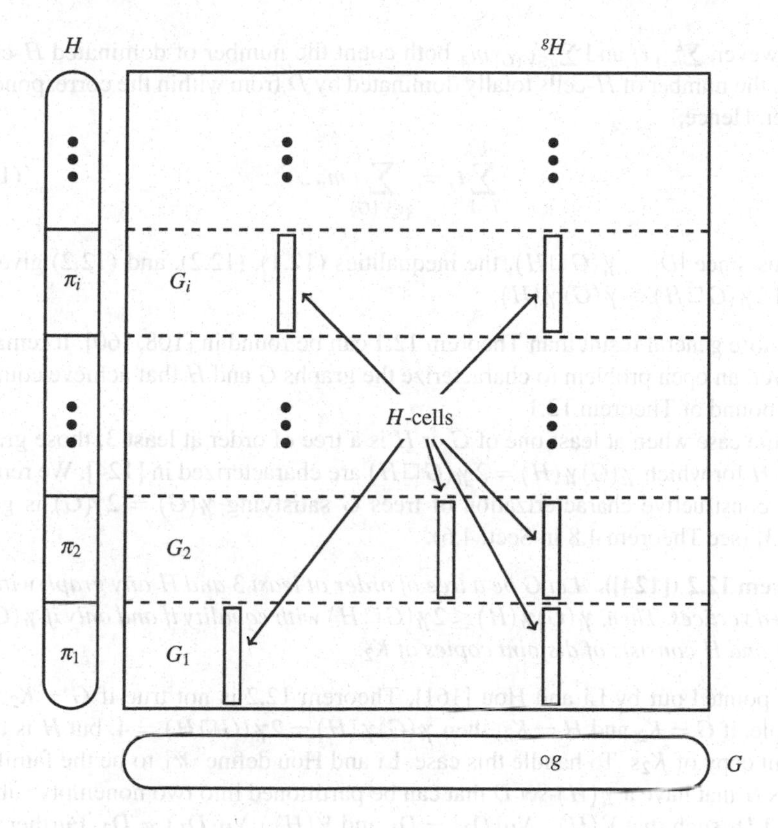

Fig. 12.1 H-cells in $G \square H$

On the other hand, consider an arbitrary H-cell $\{g\} \times \pi_j$ for some $g \in V(G)$ and some $j \in \{1,2,\ldots,k\}$. Every vertex in this H-cell is totally dominated in $G \square H$ by the vertex (g,h_j) since the vertex h_j is adjacent to every vertex in the set π_j in the graph H. For each vertex $g \in V(G)$, let m_g denote the number of H-cells $\{g\} \times \pi_j$ that are dominated H-cells. Each vertex from one of these m_g H-cells is totally dominated by the set $D \cap V(^gH)$. If an H-cell $\{g\} \times \pi_j$ is not a dominated H-cell (in the H-fiber ^gH) for some $j \in \{1,2,\ldots,k\}$, then as remarked earlier each vertex in this H-cell is totally dominated in $G \square H$ by the vertex (g,h_j). Let $\mathscr{J}_g$ denote the set of all such j for which the H-cell $\{g\} \times \pi_j$ is not a dominated H-cell and note that $|\mathscr{J}_g| = k - m_g$. By the above arguments, the projection $p_H(D \cap V(^gH))$ can be extended to a TD-set in H by adding to it the set $\cup_{j \in \mathscr{J}_g} \{h_j\}$. Consequently, $k = \gamma_t(H) \le |p_H(D \cap V(^gH))| + (k - m_g)$, or, equivalently, $m_g \le |p_H(D \cap V(^gH))|$. Hence summing over all $g \in V(G)$, we have that

$$\sum_{g \in V(G)} m_g \le \sum_{g \in V(G)} |p_H(D \cap V(^gH))| = |D|. \tag{12.2}$$

However, $\sum_{i=1}^{k} \ell_i$ and $\sum_{g \in V(G)} m_g$ both count the number of dominated H-cells, that is, the number of H-cells totally dominated by D from within the corresponding H-fiber. Hence,

$$\sum_{i=1}^{k} \ell_i = \sum_{g \in V(G)} m_g . \tag{12.3}$$

Thus since $|D| = \gamma_t(G \Box H)$, the inequalities (12.1), (12.2), and (12.2) give the bound $2\gamma_t(G \Box H) \geq \gamma_t(G)\gamma_t(H)$. □

A more general result than Theorem 12.1 can be found in [108, 160]. It remains, however, an open problem to characterize the graphs G and H that achieve equality in the bound of Theorem 12.1.

In the case when at least one of G or H is a tree of order at least 3, those graphs G and H for which $\gamma_t(G)\gamma_t(H) = 2\gamma_t(G \Box H)$ are characterized in [124]. We remark that a constructive characterization of trees G satisfying $\gamma_t(G) = 2\gamma(G)$ is given in [103] (see Theorem 4.8 in Sect. 4.6).

Theorem 12.2 ([124]). *Let G be a tree of order at least* 3 *and H any graph without isolated vertices. Then, $\gamma_t(G)\gamma_t(H) \leq 2\gamma_t(G \Box H)$ with equality if and only if $\gamma_t(G) = 2\gamma(G)$ and H consists of disjoint copies of K_2.*

As pointed out by Li and Hou [161], Theorem 12.2 is not true if $G = K_2$. For example, if $G = K_2$ and $H = K_3$, then $\gamma_t(G)\gamma_t(H) = 2\gamma_t(G \Box H) = 4$, but H is not a disjoint copy of K_2s. To handle this case, Li and Hou define $\mathscr{H}_1$ to be the family of graphs H that have a $\gamma_t(H)$-set D that can be partitioned into two nonempty subsets D_1 and D_1 such that $V(H) \setminus N_H(D_2) = D_1$ and $V(H) \setminus N_H(D_1) = D_2$. Further they define $\mathscr{H}_2$ to be the family of graphs H whose vertex set $V(H)$ can be partitioned into two nonempty subsets V_1 and V_2 such that $H_1 = H[V_1]$ satisfies $\gamma_t(H_1) = 2\gamma(H_1)$, $H_2 = H[V_2] \in \mathscr{H}_1$ and $\gamma_t(H) = \gamma_t(H_1) + \gamma_t(H_2)$.

Theorem 12.3 ([161]). *If H is any graph without isolated vertices, then $\gamma_t(K_2 \Box H) = \gamma_t(H)$ if and only if $\gamma_t(H) = 2\gamma(H)$ or $H \in \mathscr{H}_1 \cup \mathscr{H}_2$.*

Li and Hou [161] also prove the following result.

Theorem 12.4 ([161]). *Let $n \geq 3$ and let H be any graph without isolated vertices. Then, $\gamma_t(C_n)\gamma_t(H) = 2\gamma_t(C_n \Box H)$ if and only if $n \in \{3, 6\}$ and H consists of disjoint copies of K_2.*

Gravier [72] determined the total domination number of the Cartesian product $\gamma_t(P_m \Box P_n)$ where $1 \leq m \leq 4$.

Theorem 12.5 ([72]). *The following hold:*

(a) *For $n \geq 2$, $\gamma_t(P_1 \Box P_n) = \gamma_t(P_n) = \lfloor n/2 \rfloor + \lceil n/4 \rceil - \lfloor n/4 \rfloor$.*
(b) *For $n \geq 2$, $\gamma_t(P_2 \Box P_n) = 2\lfloor (n+2)/3 \rfloor$.*
(c) *For $n \geq 3$, $\gamma_t(P_3 \Box P_n) = n$.*
(d) *For $n \geq 4$, $\gamma_t(P_4 \Box P_n) = 2\lceil (3n+2)/5 \rceil$.*

Goddard [73] determined the exact value of the total domination number in $m \times n$ grids for n sufficiently large and for $5 \leq m \leq 8$. For this purpose, Goddard defined the function $\varepsilon(a,b)$ to be 1 if $n \equiv a \bmod b$ and 0 otherwise.

Theorem 12.6 ([73]). *For n sufficiently large, the following hold:*

(a) $\gamma_t(P_5 \square P_n) = \lceil (3n+2)/2 \rceil + \varepsilon(0,4)$.
(b) $\gamma_t(P_6 \square P_n) = 2\lceil (6n+6)/7 \rceil - 2\varepsilon(4,7)$.
(c) $\gamma_t(P_7 \square P_n) = 2n+2-\varepsilon(1,2)$.
(d) $\gamma_t(P_8 \square P_n) = 2\lceil (10n+10)/9 \rceil - 2\varepsilon(1,9) + 2\varepsilon(7,9)$.

Gravier [72] gave the following general bound on $\gamma_t(P_m \square P_n)$ for m and n sufficiently large.

Proposition 12.7 ([72]). *For integers $m \geq 16$ and $m \geq 16$,*

$$\frac{3mn+2n+2m}{12} - 1 \leq \gamma_t(P_n \square P_m) \leq \frac{3mn+6n+6m}{12} - 3.$$

Excellent surveys of domination in Cartesian products have been written by Hartnell and Rall [80] and Imrich and Klavžar [145].

12.3 The Direct Product

The *direct product* (also known as the *cross product*, *tensor product*, *categorical product*, and *conjunction*) of two graphs G and H, denoted by $G \times H$, is the graph with vertex set $V(G) \times V(H)$ where two vertices (u_1, v_1) and (u_2, v_2) are adjacent in $G \times H$ if and only if $u_1u_2 \in E(G)$ and $v_1v_2 \in E(H)$.

Nowakowski and Rall [169] observed that if $\otimes$ is any graph product such that the direct product $G \times H$ is a spanning subgraph of $G \otimes H$ for all graphs G and H, then the (set) Cartesian product of TDSs of G and H is a TDS of $G \otimes H$. As an immediate consequence of this result, we have the following theorem.

Theorem 12.8 ([169]). *For graphs G and H without isolated vertices, $\gamma_t(G \times H) \leq \gamma_t(G)\gamma_t(H)$.*

The inequality in Theorem 12.8 can be strict. For example, $\gamma_t(K_3 \times K_3) = 3 < 4 = \gamma_t(K_3)\gamma_t(K_3)$. Rall [175] established a family of graphs G that achieve equality in Theorem 12.8. Recall that $\rho^o(G)$ is the maximum cardinality of an open packing in G.

Theorem 12.9 ([175]). *Let G and H be graphs without isolated vertices. If $\gamma_t(G) = \rho^o(G)$, then $\gamma_t(G \times H) = \gamma_t(G)\gamma_t(H)$.*

To illustrate the sharpness of Theorem 12.9, consider the family of connected graphs for which each vertex is either an end-vertex or a support vertex. Every graph G in this family satisfies $\gamma_t(G) = \rho^o(G)$. Recall (see Theorem 4.5) that every tree T

of order at least 2 satisfies $\gamma_t(T) = \rho^o(T)$. Hence as a special case of Theorem 12.9, we have the following result.

Theorem 12.10 ([175]). *If T is any tree of order at least two and H is any graph without isolated vertices, then $\gamma_t(T \times H) = \gamma_t(T)\gamma_t(H)$.*

In particular, note that for $n \geq 2$, $\gamma_t(P_n \times H) = \gamma_t(P_n)\gamma_t(H)$. El-Zahar, Gravier, and Klobucar [60] determined the exact value of $\gamma_t(C_n \times K_m)$.

Theorem 12.11 ([60]). *For $n, m \geq 3$, $\gamma_t(C_n \times K_2) = 2\gamma_t(C_n)$ and $\gamma_t(C_n \times K_m) = n$.*

The exact value of $\gamma_t(C_n \times C_m)$ for all values of $n, m \geq 3$ is also established in [60] and can be determined from Observation 2.9 and the following result.

Theorem 12.12 ([60]). *The following hold:*

(a) *If k is even, then $\gamma_t(C_{2k} \times C_m) = k\gamma_t(C_m)$.*
(b) *If k is odd, then $\gamma_t(C_{2k} \times C_m) = 2\gamma_t(C_k \times C_m)$.*
(c) *If n and m are odd integers, where $n \geq m$, then $\gamma_t(C_k \times C_m) = \lceil n\gamma_t(C_m)/2 \rceil$.*

Further results on total domination in direct products of graphs can be found in [51].

Chapter 13
Graphs with Disjoint Total Dominating Sets

13.1 Introduction

A classical result in domination theory is that if S is a minimal dominating set of a graph $G = (V,E)$ without isolates, then $V \setminus S$ is also a dominating set of G. Thus, the vertex set of every graph without any isolates can be partitioned into two dominating sets. However it is not the case that the vertex set of every graph with at least four vertices can be partitioned into two TD-sets, even if every vertex has degree at least 2. A partition of the vertex set can also be thought of as a coloring. In particular, a partition into two TD-sets is a 2-coloring of the graph such that no vertex has a monochromatic (open) neighborhood.

The *total domatic number* of a graph G, denoted by $\mathrm{tdom}(G)$, is the maximum number of TD-sets into which the vertex set of G can be partitioned. A related problem is to determine the *domatic number* of G, denoted by $\mathrm{dom}(G)$, which is the maximum number of dominating sets into which the vertex set of G can be partitioned. Since every TD-set is a dominating set, every partition of $V(G)$ into TD-sets is a partition of $V(G)$ into dominating set, and so $\mathrm{dom}(G) \geq \mathrm{tdom}(G)$. If G is a graph with $\mathrm{dom}(G) = 2k$ for some $k \geq 1$, then combining pairs of dominating sets in the associated partition of $V(G)$ into $2k$ dominating sets partition $V(G)$ into k TD-sets, implying that $\mathrm{tdom}(G) \geq k$. Further if $\mathrm{dom}(G) = 2k+1$ for some $k \geq 1$, then $\mathrm{tdom}(G) \geq k$ since we combine three dominating sets into one set and pair of the remaining dominating sets in the partition. Hence we have the following relation between the domatic number and the total domatic number of a graph.

Theorem 13.1. *If G is a graph with no isolated vertex, then*

$$\mathrm{tdom}(G) \leq \mathrm{dom}(G) \leq 2\mathrm{tdom}(G) + 1.$$

M.A. Henning and A. Yeo, *Total Domination in Graphs*, Springer Monographs in Mathematics, DOI 10.1007/978-1-4614-6525-6_13,

We remark that equality can occur in the upper bound of Theorem 13.1. For example, if $G = C_n$, where $n \equiv 0 \pmod 3$ but $n \not\equiv 0 \pmod 4$, then $\mathrm{dom}(G) = 3$, but $\mathrm{tdom}(G) = 1$.

Zelinka [211,212] showed that no minimum degree is sufficient to guarantee the existence of a partition into two TD-sets. Consider the bipartite graph G_n^k formed by taking as one partite set, a set A of n elements, and as the other partite set all the k-element subsets of A and joining each element of A to those subsets it is a member of. Then G_n^k has minimum degree k. As observed in [211], if $n \ge 2k-1$, then in any 2-coloring of A at least k vertices must receive the same color, and these k are the neighborhood of some vertex.

In contrast to Zelinka's somewhat negative result, results of Calkin and Dankelmann [23], Feige et al. [69], and Yuster [209] show that if the maximum degree is not too large relative to the minimum degree, then sufficiently large minimum degree does suffice. For example, Calkin and Dankelmann [23] give a lower bound on the domatic number of a graph in terms of order, minimum degree, and maximum degree.

Theorem 13.2 ([23]). *Let G be a graph of order n with minimum degree δ and maximum degree Δ, and let k be a nonnegative integer. If*

$$e(\Delta^2+1)k\left(k-\frac{1}{k}\right)^{\delta+1} < 1,$$

then $\mathrm{dom}(G) \ge k$.

Yuster [209] showed that if G is a graph with minimum degree k and maximum degree at most Ck for some fixed real number $C \ge 1$, then $\mathrm{dom}(G) \ge (1 + o_k(1))k/(2\ln k)$.

Heggernes and Telle [101] showed that the decision problem to decide for a given graph G if there is a partition of $V(G)$ into two TD-sets is NP-complete, even for bipartite graphs.

13.2 Augmenting a Graph

Broere et al. [15] considered the question of how many edges must be added to G to ensure a partition of $V(G)$ into two TD-sets in the resulting graph. They denote this minimum number by $\mathrm{td}(G)$. It is clear that $\mathrm{td}(G)$ can only exist for graphs with at least four vertices. In particular, they show that:

Theorem 13.3 ([15]). *If T is a tree with ℓ leaves, then* $\ell/2 \le \mathrm{td}(T) \le \ell/2+1$.

Dorfling et al. [54] further explored this problem of augmenting a graph of minimum degree 2 to have two disjoint TD-sets.

Theorem 13.4 ([54]). *If G is a graph on $n \geq 4$ vertices with minimum degree at least* 2*, then* $\mathrm{td}(G) \leq \frac{1}{4}(n - 2\sqrt{n}) + c\log n$ *for some constant c.*

The result of Theorem 13.4 is in a sense best possible as may be seen by considering the following graph. Start with the complete graph on $2k$ vertices, duplicate each edge, and then subdivide each edge. The resultant graph G of order $n = 4k^2$ satisfies $\mathrm{td}(G) = 2\binom{k}{2} = (n - 2\sqrt{n})/4$.

13.3 Disjoint Dominating and Total Dominating Sets

In this section, we consider the question of which graphs have the property that their vertex set can be partitioned into a dominating set and a total dominating set. We shall call such a graph a DTDP-graph (standing for "dominating, total dominating, partitionable graph").

As remarked in [125], not every graph with minimum degree one is a DTDP-graph. The simplest such counterexample is a star $K_{1,n}$.

The graph obtained from the corona $\mathrm{cor}(H) = H \circ K_1$ of an arbitrary graph H by subdividing at least one of the added pendant edges is another example of a graph that is not a DTDP-graph and whose diameter can be made arbitrarily large (by choosing H to have large diameter).

In [125], the authors consider the question of whether every graph with minimum degree at least two is a DTDP-graph. Clearly the vertex set of a 5-cycle C_5 cannot be partitioned into a dominating set and a TD-set. However this is shown to be the only exception.

Theorem 13.5 ([125]). *Every graph with minimum degree at least two that contains no C_5-component is a DTDP-graph.*

As remarked earlier, the minimum degree condition of Theorem 13.5 cannot be relaxed from minimum degree two to minimum degree one.

The authors in [118] and [119] define a *DT-pair* of a graph G, if it exists, to be a pair (D,T) of disjoint sets of vertices of G such that D is a dominating set and T is a TD-set of G. The parameter $\gamma\gamma_t(G)$ is defined in [119] as follows:

$$\gamma\gamma_t(G) = \min\{|D| + |T| \colon (D,T) \text{ is DT-pair of } G\}.$$

Theorem 13.5 implies that every graph G with minimum degree at least 2 and with no C_5-component has a DT-pair, and so $\gamma\gamma_t(G)$ exists for all such graphs G. Hence we have the following immediate consequence of Theorem 13.5.

Theorem 13.6 ([125]). *If G is a graph with minimum degree at least* 2 *and with no C_5-component, then $\gamma\gamma_t(G) \leq |V(G)|$.*

In [119] and [118] graphs that achieve equality in the upper bound in Theorem 13.6 are studied. The following result [118] gives a partial answer to the question for which graphs Theorem 13.5 is best possible in the sense that the union $D \cup T$ of the two sets necessarily contains all vertices of the graph G.

Theorem 13.7 ([118]). *If G is a graph of minimum degree at least 3 with at least one component different from the Petersen graph, then $\gamma\gamma_t(G) < |V(G)|$.*

Hence if we restrict our attention to graphs with minimum degree at least 3, then a characterization of such graphs is given by Theorem 13.7 which shows that every component is the Petersen graph. However the situation becomes much more complicated when we relax the degree condition from minimum degree at least 3 to minimum degree at least 2. In this case a characterization seems difficult to obtain since there are several families each containing infinitely many graphs that achieve equality in Theorem 13.6. For example, for $k \geq 2$, if G is the connected graph that can be constructed from k disjoint 5-cycles by identifying a set of k vertices, one from each cycle, into one vertex, then $\gamma\gamma_t(G) = |V(G)|$. Several other such families are constructed in [119] which indicate that a characterization of general graphs that achieve equality in Theorem 13.6 seems difficult to obtain. The authors in [119] therefore restrict their attention to graphs with minimum degree at least two and with no induced 5-cycle and defined the families $\mathscr{C}^*$ and $\mathscr{K}^*$ of particular cycles and subdivided complete graphs as follows:

$$\mathscr{C}^* = \{C_n : n \geq 3 \text{ and } n \neq 5\}$$
$$\mathscr{K}^* = \{K_n^* : n \geq 4\},$$

where K_n^* denotes the graph obtained from a complete graph K_n, where $n \geq 4$, by subdividing every edge once. The following characterization of the C_5-free graphs which achieve equality in Theorem 13.6 is given in [119].

Theorem 13.8 ([119]). *Let G be a connected C_5-free graph with $\delta(G) \geq 2$. Then, $\gamma\gamma_t(G) = |V(G)|$ if and only if $G \in \mathscr{C}^* \cup \mathscr{K}^*$.*

A constructive characterization of DTDP-graphs is provided in [126]. Before giving this construction we will consider DTDP-trees. The key to their constructive characterization is to find a labeling of the vertices that indicates the role each vertex plays in the sets associated with both parameters. This idea of labeling the vertices is introduced in [56], where trees with equal domination and independent domination numbers as well as trees with equal domination and total domination numbers are characterized.

The labeling of a graph G is defined in [126] as a partition $S = (S_A, S_B)$ of $V(G)$. The *label* or *status* of a vertex v, denoted sta(v), is the letter $x \in \{A, B\}$ such that $v \in S_x$. The aim is to describe a procedure to build DTDP-graphs in terms of labelings. By a labeled-P_4, we shall mean a P_4 with the two central vertices labeled A and the two leaves labeled B.

Let $\mathscr{T}_{lt}$ be the minimum family of labeled trees that (i) contains a labeled-P_4 and (ii) is closed under the four operations $\mathscr{O}_1$, $\mathscr{O}_2$, $\mathscr{O}_3$, and $\mathscr{O}_4$ listed below, which extend a labeled tree T by attaching a tree to the vertex $v \in V(T)$.

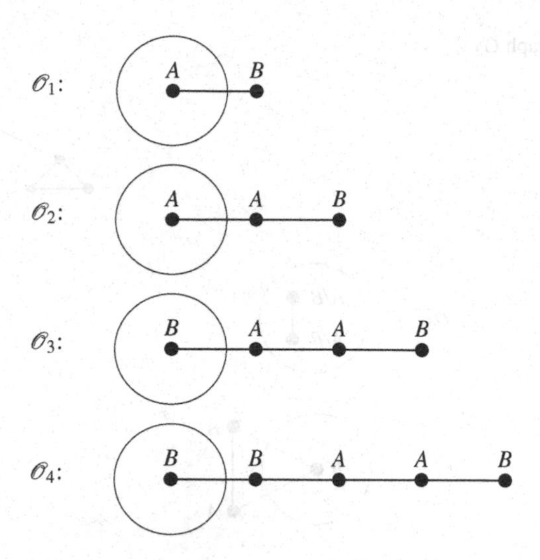

Fig. 13.1 The four operations $\mathcal{O}_1$, $\mathcal{O}_2$, $\mathcal{O}_3$, and $\mathcal{O}_4$

- *Operation* $\mathcal{O}_1$. Let v be a vertex with $\text{sta}(v) = A$. Add a vertex u_1 and the edge vu_1. Let $\text{sta}(u_1) = B$.
- *Operation* $\mathcal{O}_2$. Let v be a vertex with $\text{sta}(v) = A$. Add a path u_1u_2 and the edge vu_1. Let $\text{sta}(u_1) = A$ and $\text{sta}(u_2) = B$.
- *Operation* $\mathcal{O}_3$. Let v be a vertex with $\text{sta}(v) = B$. Add a path $u_1u_2u_3$ and the edge vu_1. Let $\text{sta}(u_1) = \text{sta}(u_2) = A$ and $\text{sta}(u_3) = B$.
- *Operation* $\mathcal{O}_4$. Let v be a vertex with $\text{sta}(v) = B$. Add a path $u_1u_2u_3u_4$ and the edge vu_1. Let $\text{sta}(u_1) = \text{sta}(u_4) = B$ and $\text{sta}(u_2) = \text{sta}(u_3) = A$.

These four operations $\mathcal{O}_1$, $\mathcal{O}_2$, $\mathcal{O}_3$, and $\mathcal{O}_4$ are illustrated in Fig. 13.1.

The following result determines which trees are DTDP-trees by establishing the following constructive characterization of DTDP-trees that uses labelings.

Theorem 13.9 ([126]). *The DTDP-trees are precisely those trees T such that $(T,S) \in \mathcal{T}_{lt}$ for some labeling S.*

As remarked in [126], if a connected graph has a spanning DTDP-tree, then it is a DTDP-graph. However, a connected DTDP-graph does not necessarily have a spanning DTDP-tree. For example, let G_k be obtained from the disjoint union of $k \geq 1$ copies of K_3 by adding a path P_3 and joining a leaf of the added path to one vertex from each copy of K_3. The graph G_3 is illustrated in Fig. 13.2. Then, G_k is a DTDP-graph, but G_k does not have a spanning DTDP-tree. We remark that we could have replaced some or all of the copies of K_3 in G_k with copies of C_6 or C_9. Hence we need a deeper result than the characterization of DTDP-trees presented in Theorem 13.9 in order to characterize DTDP-graphs in general. We will present such a characterization in Theorem 13.10.

Every DTDP-graph has order at least 3. Trivially, the only DTDP-graph of order 3 is the complete graph K_3. In order to characterize the DTDP-graphs of order

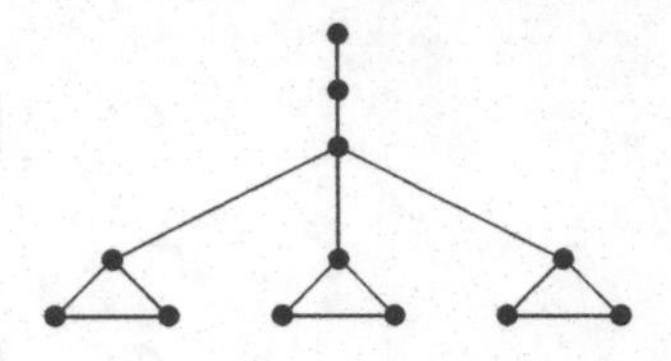

Fig. 13.2 The graph G_3

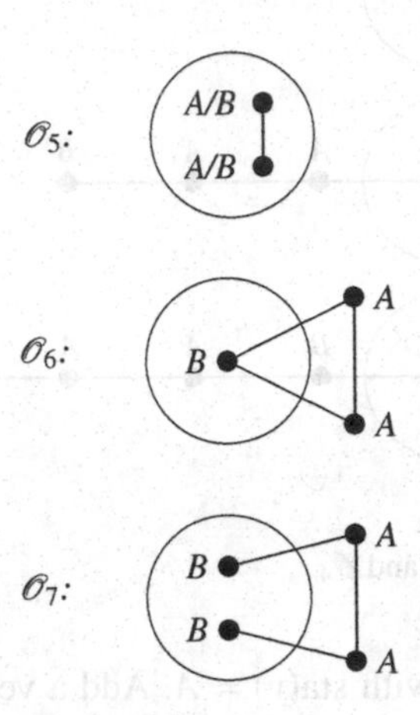

Fig. 13.3 The three operations $\mathcal{O}_5$, $\mathcal{O}_6$, and $\mathcal{O}_7$

at least 4 using labelings, the authors in [126] introduce three additional operations $\mathcal{O}_5$, $\mathcal{O}_6$, and $\mathcal{O}_7$ listed below, which extend a labeled graph G as follows:

- *Operation* $\mathcal{O}_5$. Let u and v be two nonadjacent vertices in G. Add the edge uv.
- *Operation* $\mathcal{O}_6$. Let $v \in V(G)$ and assume $\text{sta}(v) = B$. Add a path u_1u_2 and the edges vu_1 and vu_2. Let $\text{sta}(u_1) = \text{sta}(u_2) = A$.
- *Operation* $\mathcal{O}_7$. Let u and v be distinct vertices of G. Assume $\text{sta}(u) = \text{sta}(v) = B$. Add a path u_1u_2 and the edges uu_1 and vu_2. Let $\text{sta}(u_1) = \text{sta}(u_2) = A$.

These three operations are illustrated in Fig. 13.3.

Let $\mathcal{G}_{lg}$ be the minimum family of labeled graphs that (i) contains a labeled-P_4 and (ii) is closed under the seven operations $\mathcal{O}_1, \mathcal{O}_2, \ldots, \mathcal{O}_7$ described earlier. By construction, the family $\mathcal{T}_{lt}$ is a subfamily of the family $\mathcal{G}_{lg}$.

The following constructive characterization of DTDP-graphs of order at least 4 that uses labelings is presented in [126].

Theorem 13.10 ([126]). *The connected DTDP-graphs of order at least* 4 *are precisely those graphs G such that* $(G, S) \in \mathcal{G}_{lg}$ *for some labeling S.*

13.4 Dominating and Total Dominating Partitions in Cubic Graphs

Recall that by Theorem 5.7 every cubic graph on n vertices has a TD-set of cardinality at most $n/2$ and this bound is tight. In this section, we show that every connected cubic graph on n vertices has a TD-set whose complement contains a dominating set such that the cardinality of the TD-set is at most $(n+2)/2$, and this bound is essentially best possible.

Recall that a DT-pair of a graph G, if it exists, is a pair (D,T) of disjoint sets of vertices of G such that D is a dominating set and T is a TD-set of G. In [189] the authors define a *DT-pair total dominating set*, abbreviated DT-pair TD-set, to be a TD-set $T \subseteq V$ such that $V \setminus T$ contains a dominating set. Further the *DT-pair total domination number* of G, denoted by $\gamma\gamma_t^*(G)$, is defined to be the minimum cardinality of a DT-pair TD-set of G. A DT-pair TD-set of G of cardinality $\gamma\gamma_t^*(G)$ is called a $\gamma\gamma_t^*(G)$-set.

Note that $\gamma_t(G) \le \gamma\gamma_t^*(G)$, as every DT-pair TD-set of G is also a TD-set of G. There are many examples where this inequality is strict, for example, the Petersen graph G_{10} shown in Fig. 13.4. Every $\gamma_t(G_{10})$-set is of the form $N[v]$, where v is an arbitrary vertex in G_{10}, but the set $V(G_{10}) \setminus N[v]$ is not a dominating set in G_{10}. Therefore, $\gamma\gamma_t^*(G_{10}) > \gamma_t(G_{10}) = 4$. On the other hand, taking T to be the set of five vertices on the outer cycle of G_{10} (as drawn in Fig. 13.4) gives us a DT-pair TD-set of G_{10} of size five. Therefore $\gamma\gamma_t^*(G_{10}) = 5 > 4 = \gamma_t(G_{10})$.

Consider also the cubic graph P' of order $n = 20$ constructed from two copies of the Petersen graph by removing an edge from each copy and adding the two edges shown in Fig. 13.4. Then, $\gamma\gamma_t^*(P') = 9 > 8 = \gamma_t(P')$.

The following table was obtained in [189] using a computer search.

The number of non-isomorphic cubic graphs G of girth at least 5 and $\gamma_t(G) < \gamma\gamma_t^*(G)$, [189].	
$\lvert V(G)\rvert = 20$	3
$\lvert V(G)\rvert = 22$	835
$\lvert V(G)\rvert = 24$	5890

The following upper bound on the DT-pair total domination number of a connected cubic graph in terms of its order is established in [189].

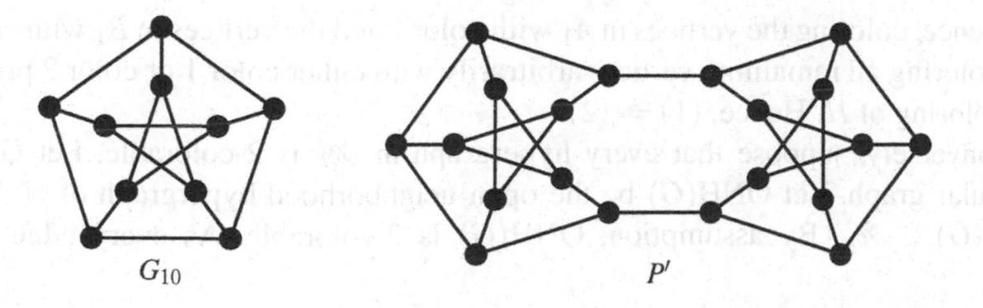

Fig. 13.4 The Petersen graph G_{10} and the constructed graph P' of order 20

Theorem 13.11 ([189]). *If G is a connected cubic graph of order n, then $\gamma\gamma_t^*(G) \leq (n+2)/2$.*

The bound of Theorem 13.11 is almost sharp since there exist two infinite families of connected cubic graphs G of order n such that $\gamma\gamma_t^*(G) = n/2$ as shown in the following result.

Proposition 13.12 ([189]). *If $G \in \mathscr{G} \cup \mathscr{H} \cup \{G_{16}\} \cup \{G_{10}\}$ has order n, where $\mathscr{G}$ and $\mathscr{H}$ are the two families constructed in Sect. 5.5, G_{16} is the generalized Petersen graph shown in Fig. 1.2 and G_{10} is the Petersen graph shown in Fig. 13.4, then $\gamma\gamma_t^*(G) = n/2$.*

13.5 A Surprising Relationship

In this section, we exhibit a surprising connection between disjoint total dominating sets in graphs, 2-coloring of hypergraphs, and even cycles in digraphs. For $k \geq 2$, let $\mathscr{H}_k$ denote the class of all k-uniform k-regular hypergraphs, where we recall that a hypergraph H is k-regular if every vertex has degree k in H. We also recall König's Theorem [153] that every regular bipartite graph contain a perfect matching. We remark that using results on permanents and determinants of the adjacency matrix, it is proved in [4] that Part (4) in Theorem 13.13 holds for all $k \geq 8$. Then (4) $\Rightarrow$ (2) was proved in [4]. The proof of Theorem 13.13 we present here can be found in [141].

Theorem 13.13 ([141]). *The following statements are equivalent for all $k \geq 1$:*

(1) *Every k-regular graph contains two disjoint total dominating sets.*
(2) *Every hypergraph in $\mathscr{H}_k$ is 2-colorable.*
(3) *Every $(k-1)$-regular digraph has an even cycle.*
(4) *Every k-regular bipartite graph, G, contains a cycle, C, such that $|V(C)| = 0 \pmod 4$ and $G - V(C)$ contain a perfect matching.*

Proof. Suppose that every k-regular graph contains two disjoint total dominating sets. Let $H = (V,E)$ be a hypergraph in $\mathscr{H}_k$ and let G_H be the incidence bipartite graph of H with partite sets V and E. Then, G_H is a k-regular bipartite graph. By assumption, G_H contains two disjoint total dominating sets A and B. Let $A_1 = A \cap V$ and let $B_1 = B \cap V$. Then every hyperedge in H contains a vertex from both A_1 and B_1. Hence, coloring the vertices in A_1 with color 1 and the vertices in B_1 with color 2 and coloring all remaining vertices arbitrarily with either color 1 or color 2 produce a 2-coloring of H. Hence, (1) $\Rightarrow$ (2).

Conversely, suppose that every hypergraph in $\mathscr{H}_k$ is 2-colorable. Let G be a k-regular graph. Let ONH(G) be the open neighborhood hypergraph of G. Then, ONH(G) $\in \mathscr{H}_k$. By assumption, ONH(G) is 2-colorable. As every edge of H

contains vertices of both colors, every open neighborhood in G contains vertices of both colors. Thus the two color classes in H form two disjoint total dominating sets in G. Thus, (2) $\Rightarrow$ (1).

Suppose that every hypergraph in $\mathscr{H}_k$ is bipartite. Let $D = (V,A)$ be a $(k-1)$-regular digraph. For each vertex $v \in V$, let e_v be the set $N^+[v]$ that consists of v and its $k-1$ out-neighbors. Let H be the hypergraph with vertex set V and with edge set $E = \{e_v \mid v \in V\}$. Then, $H \in \mathscr{H}_k$ and is therefore 2-colorable. Let D^* be the subdigraph of D obtained by deleting all arcs where the two endpoints have the same color. As H was 2-colored we note that $\delta^+(D^*) \geq 1$, which implies that D^* contains a cycle. By construction this cycle is even and belongs to D. Thus, (2) $\Rightarrow$ (3).

Suppose that every $(k-1)$-regular digraph has an even cycle and let G be a k-regular bipartite graph. By König's Theorem there exists a perfect matching M in G. Without loss of generality assume that $M = \{a_1b_1, a_2b_2, \ldots, a_rb_r\}$ where $\{a_1, a_2, \ldots, a_r\}$ and $\{b_1, b_2, \ldots, b_r\}$ are the partite sets of G. Construct the $(k-1)$-regular digraph D such that $V(D) = \{x_1, x_2, \ldots, x_r\}$ and $E(D) = \{(x_i, x_j) \mid b_ia_j \in E(G) \text{ and } i \neq j\}$. By assumption, D has an even cycle, C, and without loss of generality $C = x_1x_2 \ldots x_{2s}x_1$. However now $a_1b_1a_2b_2a_3 \ldots a_{2s}b_{2s}a_1$ is a cycle of length $0 \pmod 4$ and $a_{2s+1}b_{2s+1}, a_{2s+2}b_{2s+2}, \ldots, a_rb_r$ is a perfect matching in $G - M$. Thus, (3) $\Rightarrow$ (4).

It remains for us to show that (4) $\Rightarrow$ (2), which was proved in [4]. Suppose that every k-regular bipartite graph, G, contains a cycle, C, such that $|V(C)| = 0 \pmod 4$ and $G - V(C)$ contain a perfect matching and let $H \in \mathscr{H}_k$ have order r. We may assume that H is connected since a hypergraph is 2-colorable if and only if each component is 2-colorable. Let G_H be the incidence bipartite graph of H with partite sets $V = V(H)$ and $E = E(H)$. Then, G_H is a k-regular bipartite graph and therefore contains a cycle C, such that $|V(C)| = 0 \pmod 4$ and $G - V(C)$ contains a perfect matching, M. Without loss of generality we may assume that $C = v_1e_1v_2e_2 \ldots v_{2s}e_{2s}v_1$ and $M = \{v_{2s+1}e_{2s+1}, v_{2s+2}e_{2s+2}, \ldots, v_re_r\}$ where $v_i \in V(H)$ and $e_i \in E(H)$ for all $i = 1, 2, \ldots, r$. Assign color 1 to $v_1, v_3, v_5, \ldots, v_{2s-1}$ and assign color 2 to $v_2, v_4, \ldots, v_{2s}$ and let $I = \{1, 2, \ldots, 2s\}$ denote the indices of the vertices in H which have been assigned colors. Note that e_i contains vertices of different color for all $i \in I$.

Let $J = \{1, 2, \ldots, n\} \setminus I$. If $J = \emptyset$, then we are done as every edge in H contains vertices of different color and we therefore have a 2-coloring of H. So assume that $J \neq \emptyset$. If no edge e_j contains a vertex v_i for any $j \in J$ and $i \in I$, then all $|J|$ edges with an index in J only contain vertices with indices in J which implies that the subgraph of G_H induced by the vertices and edges with an index in J is k-regular. This in turn implies that H is disconnected, a contradiction. Thus there exists an edge e_j containing a vertex v_i for some $j \in J$ and $i \in I$. Now add j to I and color v_j with the opposite color of v_i. We note that this operation has increased the size of I by one and that it is still true that e_i contains vertices of different color for all $i \in I$. We can therefore continue the above process until all vertices in H have been colored in such a way that each edge of H contain vertices of both colors. Hence, (4) $\Rightarrow$ (2). We have therefore shown that (1) $\Leftrightarrow$ (2) $\Leftrightarrow$ (3) $\Leftrightarrow$ (4). $\square$

We shall need the following result due to Thomassen [198].

Theorem 13.14 ([198]). *Every k-regular digraph with $k \geq 3$ contains an even cycle.*

As an immediate consequence of Theorem 13.13 and Thomassen's result in Theorem 13.14, we have the following result.

Theorem 13.15 ([141]). *The following hold:*

(a) *Every k-regular graph contains two disjoint total dominating sets, provided $k \geq 4$.*

(b) *Every hypergraph in $\mathscr{H}_k$ is 2-colorable, provided $k \geq 4$.*

Theorem 13.15 can be rephrased in terms of transversals: Every hypergraph in $\mathscr{H}_k$ has two vertex disjoint transversals, provided $k \geq 4$. As remarked by Alon and Bregman [4], the result of Theorem 13.15 is not true when $k = 3$, as may be seen by considering the Fano plane shown in Fig. 5.10. There are infinitely many examples showing that Theorem 13.15 does not hold for $k = 3$.

Chapter 14
Total Domination in Graphs with Diameter Two

14.1 Introduction

As shown in Theorem 6.3, the total domination number of a diameter-2 planar graph is at most 3. However there exist diameter-2 nonplanar graphs G of arbitrarily large-order n such that $\gamma_t(G) \geq \sqrt{n}$. For example, for $t \geq 2$, the Cartesian product $G = K_t \,\square\, K_t$ is a diameter-2 graph of order $n = t^2$ satisfying $\gamma_t(G) = t = \sqrt{n}$. In this chapter we determine an upper bound of $1 + \sqrt{n \ln(n)}$ on the total domination number of a diameter-2 graph and show that this bound is close to optimum.

14.2 Preliminary Observations

For diameter-2 graphs of small order, it is a simple exercise to compute their total domination number (which can also be readily checked by computer).

Observation 14.1 ([50]). *If G is a diameter-2 graph of order $n \leq 11$, then $\gamma_t(G) \leq 1 + \sqrt{n}$ with equality if and only if $G = G_9$, where G_9 is the diameter-2 graph of order $n = 9$ shown in Fig. 14.1.*

We observe that if G is a diameter-2 graph of order n and v is an arbitrary vertex in G, then the diameter-2 constraint implies that $N[v]$ is a TD-set in G. Choosing v to be a vertex of minimum degree we have that $|N[v]| = \delta(G) + 1$. Hence we have the following observation.

Observation 14.2 ([50]). *If G is a diameter-2 graph of order n, then $\gamma_t(G) \leq 1 + \delta(G)$.*

Observation 14.2 implies the following result.

Corollary 14.3. *If G is a diameter-2 graph of order n with minimum degree δ, where $\delta \leq \sqrt{n}$, then $\gamma_t(G) \leq 1 + \sqrt{n}$.*

M.A. Henning and A. Yeo, *Total Domination in Graphs*, Springer Monographs in Mathematics, DOI 10.1007/978-1-4614-6525-6_14,

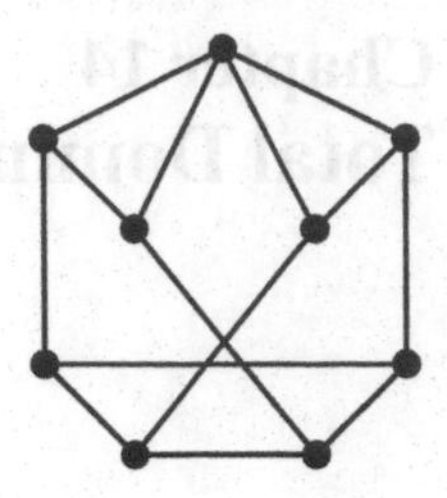

Fig. 14.1 The graph G_9 of order n with $\gamma_t(G) = 1 + \sqrt{n} = 4$

As a consequence of Theorem 5.1 in Sect. 5.2, we establish next an upper bound on the total domination number of a general graph with relative large minimum degree.

Theorem 14.4 ([50]). *If G is a graph of order $n \geq 2$ with $\delta(G) \geq \sqrt{n}\ln(n)$, then $\gamma_t(G) \leq 1 + \sqrt{n}$.*

Proof. If $n = 2$, then $\gamma_t(G) = 2 < 1 + \sqrt{n}$. Hence we may assume that $n \geq 3$. Therefore since $\delta \geq \sqrt{n}\ln(n)$, we note that $\delta \geq 2$. Applying Theorem 5.1 to the graph G and noting that $(1 + \ln(\delta))/\delta$ is a decreasing function in δ, we have that

$$\begin{aligned}
\gamma_t(G) &\leq \left(\frac{1 + \ln(\delta)}{\delta}\right) n \\
&\leq \left(\frac{1 + \ln(\sqrt{n}\ln(n))}{\sqrt{n}\ln(n)}\right) n \\
&= \left(\frac{1 + \frac{1}{2}\ln(n) + \ln(\ln(n))}{\ln(n)}\right) \sqrt{n} \\
&= \frac{\sqrt{n}}{2} + \left(\frac{1 + \ln(\ln(n))}{\ln(n)}\right) \sqrt{n}.
\end{aligned}$$

To prove that $\gamma_t(G) \leq 1 + \sqrt{n}$, it therefore suffices for us to show that

$$\left(\frac{1 + \ln(\ln(n))}{\ln(n)}\right) \sqrt{n} \leq 1 + \frac{\sqrt{n}}{2}.$$

This is clearly true when $n \geq 213$ as $(1 + \ln(\ln(n)))/(\ln(n))$ is a decreasing function, and when $n \geq 213$ it is below $1/2$. It is easy to check that it is also true for all $3 \leq n \leq 212$ (by using a computer). □

As a special case of Theorem 14.4, we have the following result.

Corollary 14.5. *If G is a diameter-2 graph of order n with minimum degree δ, where $\delta \geq \sqrt{n}\ln(n)$, then $\gamma_t(G) \leq 1 + \sqrt{n}$.*

As observed earlier, there exist diameter-2 graphs G of arbitrarily large-order n such that $\gamma_t(G) = \sqrt{n}$. In view of Corollaries 14.3 and 14.5, it is a natural question to ask whether $1 + \sqrt{n}$ is an upper bound on the total domination number of a diameter-2 graph of order n. As a consequence of these two corollaries, we note that if there do exist diameter-2 graphs G of order n for which $\gamma_t(G) > 1+\sqrt{n}$, then there is a narrow window range for the minimum degree of such graphs.

Observation 14.6. *If G is a diameter-2 graph of order n with minimum degree δ such that $\gamma_t(G) > 1+\sqrt{n}$, then $\sqrt{n} < \delta < \sqrt{n}\ln(n)$.*

In Sect. 14.4 we will show that there do exist diameter-2 graphs with $\gamma_t(G) > 1+\sqrt{n}$ and that the correct bound is of the order $1+\sqrt{n\ln(n)}$.

14.3 Diameter-2 Moore Graphs

If G is a diameter-2 graph, then G is either a star or G has girth 3, 4, or 5. The diameter-2 graphs of girth 5 are precisely the diameter-2 Moore graphs. It is shown (see [143, 185]) that Moore graphs are r-regular and that diameter-2 Moore graphs have order $n = r^2+1$ and exist for $r = 2,3,7$, and possibly 57, but for no other degrees. The Moore graphs for the first three values of r are unique, namely

- The 5-cycle (2-regular graph on $n = 5$ vertices)
- The Petersen graph (3-regular graph on $n = 10$ vertices)
- The Hoffman–Singleton graph (7-regular on $n = 50$ vertices).

The following result determines the total domination number of a diameter-2 Moore graph.

Theorem 14.7 ([50]). *If G is a diameter-2 graph of order n with girth 5, then $\gamma_t(G) = 1+\sqrt{n-1}$.*

Proof. As observer earlier, the graph $G = (V,E)$ is a diameter-2 Moore graph and $n = r^2+1$. Hence, $r = \sqrt{n-1}$. Let D be a $\gamma_t(G)$-set. Then, $V = \cup_{v\in D} N(v)$, implying that $|V| \le \sum_{v\in D} d_G(v) \le \Delta(G)\cdot|D|$, or equivalently, $\gamma_t(G) = |D| \ge |V|/\Delta(G) = n/\sqrt{n-1}$. Therefore by Observation 14.2, we have that $n/\sqrt{n-1} \le \gamma_t(G) \le 1+\sqrt{n-1}$, or, equivalently, $\sqrt{n-1}+1/\sqrt{n-1} \le \gamma_t(G) \le 1+\sqrt{n-1}$. Since both $\gamma_t(G)$ and $\sqrt{n-1}$ are integers, $\gamma_t(G) = 1+\sqrt{n-1}$. □

14.4 An Optimal Upper Bound

Recall that by Observation 14.6, if G is a diameter-2 graph of order n with minimum degree δ such that $\gamma_t(G) > 1+\sqrt{n}$, then $\sqrt{n} < \delta < \sqrt{n}\ln(n)$. It is therefore a natural question to ask whether $1+\sqrt{n}$ is an upper bound on the total domination number of a diameter-2 graph of order n. Using Theorem 5.1 we show next that $1+\sqrt{n\ln(n)}$ is an upper bound on the total domination number of a diameter-2 graph of order n.

Theorem 14.8 ([50]). *If G is a diameter-2 graph of order n, then $\gamma_t(G) \le 1 + \sqrt{n\ln(n)}$.*

Proof. If $\delta \le \sqrt{n\ln(n)}$, then $\gamma_t(G) \le 1 + \sqrt{n\ln(n)}$ by Observation 14.2. Hence we may assume that $\delta > \sqrt{n\ln(n)}$. In particular, we note that $\delta \ge 2$. Applying Theorem 5.1 to the graph G and noting that $(1+\ln(\delta))/\delta$ is a decreasing function, we have that

$$\begin{aligned}
\gamma_t(G) &\le \left(\frac{1+\ln(\delta)}{\delta}\right) n \\
&\le \left(\frac{1+\ln(\sqrt{n\ln(n)})}{\sqrt{n\ln(n)}}\right) n \\
&= \left(\frac{1+\frac{1}{2}\ln(n)+\frac{1}{2}\ln(\ln(n))}{\sqrt{\ln(n)}}\right)\sqrt{n} \\
&= \frac{\sqrt{n\ln(n)}}{2} + \left(\frac{1+\frac{1}{2}\ln(\ln(n))}{\ln(n)}\right)\sqrt{n\ln(n)}.
\end{aligned}$$

To prove that $\gamma_t(G) \le 1 + \sqrt{n\ln(n)}$, it therefore suffices for us to show that

$$\left(\frac{1+\frac{1}{2}\ln(\ln(n))}{\ln(n)}\right)\sqrt{n\ln(n)} \le 1 + \frac{\sqrt{n\ln(n)}}{2}.$$

This is clearly true when $n \ge 24$ as $(1+\frac{1}{2}\ln(\ln(n)))/(\ln(n))$ is a decreasing function, and when $n \ge 24$ it is below $1/2$. It is easy to check that it is also true for all $3 \le n \le 23$. □

Do there exist diameter-2 graphs of arbitrarily large-order n such that $\gamma_t(G) > 1+\sqrt{n}$? It is shown in [50] that this is indeed the case by proving that the upper bound in Theorem 14.8 is in a sense optimal. A key result in [50] is the following result on the existence of intersecting uniform hypergraphs with large transversal number, where we recall that a hypergraph is called an intersecting hypergraph if every two distinct edges of H have a nonempty intersection. We state this result without proof.

Theorem 14.9 ([50]). *For any constant $c > 2$, there exists a constant N_c such that for all $n \ge N_c$ there exists an intersecting uniform hypergraph H with $|V(H)| = |E(H)| = n$ and $\tau(H) > \sqrt{\frac{n\ln(n)}{4c}}$.*

Using an interplay between total domination in graphs and transversals in hypergraphs we are now able to establish the following graph-theoretic result.

Theorem 14.10 ([50]). *For any constant $c > 16$, there exist diameter-2 graphs G of all sufficiently large even orders n satisfying $\gamma_t(G) > \sqrt{\frac{1}{c} n \ln(\frac{n}{2})}$.*

Proof. Let $c > 16$ be an arbitrary given constant. It suffices for us to show that there exists a constant N_c^* such that for all even $n \geq N_c^*$ there exists a diameter-2 graph G of order n with $\gamma_t(G) > \sqrt{\frac{1}{c} n \ln(\frac{n}{2})}$.

Let $d = c/8$, and so $d > 2$. By Theorem 14.9, there exists a constant N_d such that for all $n^* \geq N_d$ there exists an intersecting k-uniform hypergraph H on n^* vertices with $\tau(H) > \sqrt{\frac{n^* \ln(n^*)}{4d}}$. Let $N_c^* = 2N_d$ and let $n \geq N_c^*$ be an arbitrary even integer. Let $n^* = n/2$, and so $n^* \geq N_d$. By Theorem 14.9, let H be an intersecting k-uniform hypergraph on n^* vertices and n^* edges with $\tau(H) > \sqrt{\frac{n^* \ln(n^*)}{4d}}$.

Consider the incidence bipartite graph $G_H(X,Y)$ of the hypergraph H with partite sets $X = V(H)$ and $Y = E(H)$ and where there is an edge between $x \in X$ and $y \in Y$ if and only if x belongs to the hyperedge y in H. Let G be obtained from $G_H(X,Y)$ by adding all edges joining vertices in X. We note that Y is an independent set in G. Further, G has order $n = 2n^*$.

We first show that G has diameter two. If $x, y \in X$, then $d_G(x,y) = 1$ since $G[X]$ is a clique. Suppose $x \in X$ and $y \in Y$. If x belongs to the edge y in H, then $d_G(x,y) = 1$. Otherwise, let x' be any vertex in the edge y in H. Since $G[X]$ is a clique, $xx' \in E(G)$, implying that $d_G(x,y) = 2$. Finally suppose that $x, y \in Y$, and so x and y are both hyperedges in H. Since H is intersecting, there is a vertex u in H that belongs to both hyperedges x and y. Thus in the graph G, we have that $u \in X$ and that u is adjacent to both x and y in G. This implies that xuy is a path in G, and so $d_G(x,y) = 2$. Therefore, G is a diameter-2 graph.

It remains for us to provide a lower bound on $\gamma_t(G)$. Let S be a $\gamma_t(G)$-set. Since Y is an independent set, the set $S \cap X$ totally dominates the set Y in G. By construction of the graph G this implies that $S \cap V(H)$ is a transversal in H. Therefore, $\gamma_t(G) = |S| \geq |S \cap X| = |S \cap V(H)| \geq \tau(H)$. Hence since $\tau(H) > \sqrt{\frac{n^* \ln(n^*)}{4d}}$ and since $|V(G)| = n = 2n^*$, we have that

$$\gamma_t(G) \geq \sqrt{\frac{n^* \ln(n^*)}{4d}} = \sqrt{\frac{\frac{n}{2} \ln(\frac{n}{2})}{\frac{c}{2}}} = \sqrt{\frac{1}{c} n \ln\left(\frac{n}{2}\right)}.$$

This completes the proof of Theorem 14.10. □

We are now in a position to prove that the upper bound in Theorem 14.8 is optimal in the following sense.

Theorem 14.11. *Given any $\varepsilon > 0$, there exist diameter-2 graphs G of all sufficiently large even orders n such that $\gamma_t(G) > (\frac{1}{4+\varepsilon})\sqrt{n \ln(n)}$.*

Proof. Given $\varepsilon > 0$, let c be any fixed number such that $16 < c < (4+\varepsilon)^2$ and note that the following holds when n is large enough (as the left-hand side tends to infinity when n gets large and the right-hand side is a constant): $[(4+\varepsilon)^2 - c]\ln(n) > (4+\varepsilon)^2\ln(2)$. Therefore the following hold:

$$
\begin{aligned}
&[(4+\varepsilon)^2 - c]\ln(n) > (4+\varepsilon)^2\ln(2) \\
\Updownarrow\; &(4+\varepsilon)^2[\ln(n) - \ln(2)] > c\ln(n) \\
\Updownarrow\; &\tfrac{1}{c}\ln(\tfrac{n}{2}) > \tfrac{1}{(4+\varepsilon)^2}\ln(n) \\
\Updownarrow\; &\sqrt{\tfrac{1}{c}\, n\ln(\tfrac{n}{2})} > (\tfrac{1}{4+\varepsilon})\sqrt{n\ln(n)}.
\end{aligned}
$$

The desired result now follows from Theorem 14.10. □

Chapter 15
Nordhaus–Gaddum Bounds for Total Domination

15.1 Introduction

In 1956 the original paper [168] by Nordhaus and Gaddum appeared. In it they gave sharp bounds on the sum and product of the chromatic numbers of a graph and its complement. Since then such results have been given for several parameters. For an excellent survey of Nordhaus–Gaddum-type relations, we refer the reader to the survey paper by Aouchiche and Hansen [6]. An overview of Nordhaus–Gaddum inequalities for domination-related parameters can be found in Chap. 10 in the domination book by Haynes et al. [85]. In this chapter, we consider Nordhaus–Gaddum inequalities involving the total domination number.

15.2 Nordhaus–Gaddum Bounds for Total Domination

In the introductory paper on total domination, Cockayne, Dawes, and Hedetniemi [37] proved a Nordhaus–Gaddum bound for the sum of the total domination numbers of a graph and its complement, while a bound for their product is given in [110].

Theorem 15.1 ([37, 110]). *Let G be a graph of order n such that neither G nor $\overline{G}$ contains isolated vertices. Then the following hold:*

(a) $\gamma_t(G) + \gamma_t(\overline{G}) \le n + 2$.
(b) $\gamma_t(G)\gamma_t(\overline{G}) \le 2n$.

In both inequalities, equality holds if and only if G or $\overline{G}$ consists of disjoint copies of K_2.

For a graph G, we let $\delta^*(G) = \min\{\delta(G), \delta(\overline{G})\}$ and we note that $|V(G)| \ge 2\delta^*(G) + 1$. Further we let $\gamma_t^*(G) = \min\{\gamma_t(G), \gamma_t(\overline{G})\}$. Favaron, Karami, and

M.A. Henning and A. Yeo, *Total Domination in Graphs*, Springer Monographs in Mathematics, DOI 10.1007/978-1-4614-6525-6_15,

Sheikholeslami [67] improved the upper bound on the sum by imposing lower bounds on the total domination numbers of a graph and its complement.

Theorem 15.2 ([67]). *Let G be a graph of order $n \geq 4$. Then the following hold:*

(a) *If $\gamma_t^*(G) \geq 3$, then $\gamma_t(G) + \gamma_t(\overline{G}) \leq \delta^*(G) + 4$.*
(b) *If $\gamma_t^*(G) \geq 4$, then $\gamma_t(G) + \gamma_t(\overline{G}) \leq \delta^*(G) + 3$.*
(c) *If $\gamma_t^*(G) \geq 4$ and $\gamma_t(G) + \gamma_t(\overline{G}) = \delta^*(G) + 3$, then $\gamma_t(G) = \gamma_t(\overline{G}) = 4$.*

Favaron et al. [67] also established the following upper bounds on the sum $\gamma_t(G) + \gamma_t(\overline{G})$ of a graph in terms of the order of the graph. This bound follows from Theorem 15.2 as $\delta^*(G) \leq (n-1)/2$. We note that Theorem 15.3 improves Theorem 15.1 part (a) when $\gamma_t^*(G) \geq 3$.

Theorem 15.3 ([67]). *If G is a graph of order $n \geq 4$ satisfying $\gamma_t^*(G) \geq 3$, then $\gamma_t(G) + \gamma_t(\overline{G}) \leq \lfloor (n+7)/2 \rfloor$.*

We observe that if G is a graph satisfying $\gamma_t^*(G) \geq 3$, then both G and its complement $\overline{G}$ have diameter two. Hence applying Theorem 14.8 to G and $\overline{G}$, we have the following improvement of Theorem 15.3.

Theorem 15.4. *If G is a graph of order n satisfying $\gamma_t^*(G) \geq 3$, then $\gamma_t(G) + \gamma_t(\overline{G}) \leq 2 + 2\sqrt{n \ln(n)}$.*

With some additional work the upper bound of Theorem 15.4 can in fact be improved slightly to $4 + \sqrt{2n \ln(n)}$. We omit the details. However the upper bound of Theorem 15.4 is close to being optimal as the following result shows.

Theorem 15.5. *Given any $\varepsilon > 0$, there exist graphs G with $\gamma_t^*(G) \geq 3$ of all sufficiently large even orders n such that $\gamma_t(G) + \gamma_t(\overline{G}) \geq (\frac{1}{4+\varepsilon})\sqrt{n \ln(n)} + 3$.*

Proof. By Theorem 14.11, there exist diameter-2 graphs G of all sufficiently large even orders n such that $\gamma_t(G) > (\frac{1}{4+\varepsilon})\sqrt{n \ln(n)}$. Since n can be chosen sufficiently large, we may assume that $\gamma_t(G) \geq 3$. We will now show that $\gamma_t(\overline{G}) \geq 3$. Assume this is not the case and let $\{u,v\}$ be a TD-set in $\overline{G}$. This implies that $uv \in E(\overline{G})$ and therefore that $uv \notin E(G)$. However as $\text{diam}(G) = 2$, there is some vertex $w \in V(G)$ such that uwv is a path in G. But then the vertex w is not totally dominated by $\{u,v\}$ in $\overline{G}$, a contradiction. Therefore, $\gamma_t(\overline{G}) \geq 3$, implying that $\gamma_t^*(G) \geq 3$ and that $\gamma_t(G) + \gamma_t(\overline{G}) \geq (\frac{1}{4+\varepsilon})\sqrt{n \ln(n)} + 3$. □

Let F_6 be the graph shown in Fig. 15.1 of order 6 obtained from a 5-cycle by adding a new vertex and joining it to three consecutive vertices on the cycle. We remark that the graph F_6 is self-complementary.

Theorem 15.6 ([67]). *If G is a graph of order $n \geq 4$ different from the cycle C_5 and such that each component of G and $\overline{G}$ has order at least 3, then*

$$\gamma_t(G) + \gamma_t(\overline{G}) \leq \frac{2n}{3} + 2,$$

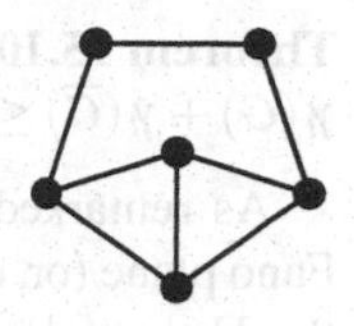

Fig. 15.1 The graph F_6

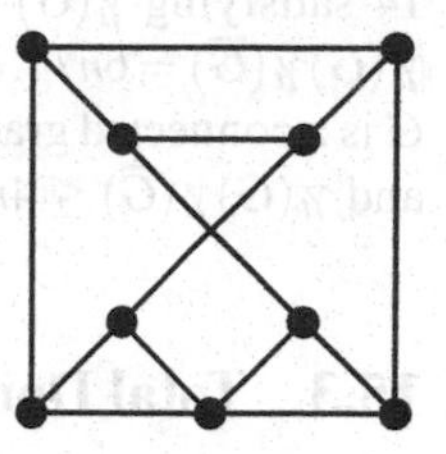

Fig. 15.2 The graph G_9

with equality if and only if G is the graph F_6 shown in Fig. 15.1 or has each of its components isomorphic to C_3 or C_6 or to the 2-corona $H \circ P_2$ of some graph H.

The upper bound in Theorem 15.1 on the product can be improved if we restrict the minimum degree on both G and $\overline{G}$ to be at least two. We remark that Theorem 15.6 implies the first part of Theorem 15.7.

Theorem 15.7 ([110]). *If G is a graph of order n such that $\delta^*(G) = 2$, then $\gamma_t(G) + \gamma_t(\overline{G}) \le 2n/3 + 2$ unless $G = C_5$ and $\gamma_t(G)\gamma_t(\overline{G}) \le 4n/3$ unless $G \in \{C_5, F_6\}$.*

As an immediate consequence of Theorem 15.7, we have the following result.

Corollary 15.8 ([110]). *Let G be a graph of order n such that $\delta^*(G) = 2$. If $n \ge 6$, then $\gamma_t(G) + \gamma_t(\overline{G}) \le 2n/3 + 2$, while if $n \ge 7$, then $\gamma_t(G)\gamma_t(\overline{G}) \le 4n/3$.*

We remark that the bounds in Theorem 15.7 and Corollary 15.8 are sharp as may be seen by taking $G = kK_3$ where $k \ge 2$ an integer. Let G_9 be the graph shown in Fig. 15.2.

Theorem 15.9 ([110]). *If G is a graph of order n such that $\delta^*(G) = 3$. Then the following hold:*

(a) $\gamma_t(G) + \gamma_t(\overline{G}) \le n/2 + 2$ *unless $G = G_9$ in which case $n = 9$ and* $\gamma_t(G) + \gamma_t(\overline{G}) = (n+5)/2$.
(b) $\gamma_t(G)\gamma_t(\overline{G}) \le n$ *for $n \ge 12$.*

Taking G to be any graph of order n different from K_4 that belongs to one of the two infinite families in Theorem 5.10, we have that $\gamma_t(G) = n/2$ and $\gamma_t(\overline{G}) = 2$. Hence the bounds in Theorem 15.9 are sharp.

Theorem 15.10 ([110]). *If G is a graph of order n such that $\delta^*(G) = 4$, then $\gamma_t(G) + \gamma_t(\overline{G}) \le 3n/7 + 2$ for $n \ge 14$, while $\gamma_t(G)\gamma_t(\overline{G}) \le 6n/7$ for $n \ge 18$.*

As remarked in [110], the incidence bipartite graph of the complement of the Fano plane (or, equivalently, the bipartite complement of the Heawood graph, where the Heawood graph is shown in Fig. 5.11) is a 4-regular graph G of order $n = 14$ satisfying $\gamma_t(G) = 3n/7$ and $\gamma_t(\overline{G}) = 2$, whence $\gamma_t(G) + \gamma_t(\overline{G}) = 3n/7 + 2$ and $\gamma_t(G)\gamma_t(\overline{G}) = 6n/7$. However, for sufficiently large n, it is conjectured in [110] that if G is a connected graph of order n such that $\delta^*(G) = 4$, then $\gamma_t(G) + \gamma_t(\overline{G}) \le 2n/5 + 2$ and $\gamma_t(G)\gamma_t(\overline{G}) \le 4n/5$.

15.3 Total Domination Number and Relative Complement

In this section, we look at another variation on Nordhaus–Gaddum type results in which we extended the concept by considering $G_1 \oplus G_2 = K(s,s)$ rather than $G_1 \oplus G_2 = K_n$. Recall that if G and H are graphs on the same vertex set but with disjoint edge sets, then $G \oplus H$ denotes the graph whose edge set is the union of their edge sets.

In [77], the authors determined the graphs H with respect to which complements are always unique in the following sense: If G_1 and G_2 are isomorphic subgraphs of H, then their complements $H - E(G_1)$ and $H - E(G_2)$ are isomorphic.

Theorem 15.11 ([77]). *Let H be a graph without isolated vertices with respect to which complements are always unique. Then H is one of the following: (a) $rK(1,s)$, (b) rK_3, (c) K_s, (d) C_5, or (e) $K(s,s)$, for some integers r and/or s.*

Theorem 15.11 suggests that the complete bipartite graph $K(s,s)$ is an obvious alternate to K_n in Nordhaus–Gaddum results. In this section we consider bounds on the sums and products of the total domination numbers of G_1 and G_2, where $G_1 \oplus G_2 = K(s,s)$ and neither G_1 nor G_2 contains isolated vertices.

Theorem 15.12 ([111]). *If $G_1 \oplus G_2 = K(s,s)$ where neither G_1 nor G_2 have an isolated vertex, then $\gamma_t(G_1) + \gamma_t(G_2) \le 2s + 4$, with equality if and only if $G_1 = sK_2$ or $G_2 = sK_2$.*

Theorem 15.13 ([111]). *If $G_1 \oplus G_2 = K(s,s)$ where neither G_1 nor G_2 have an isolated vertex, then $\gamma_t(G_1)\gamma_t(G_2) \le \max\{8s, \lfloor (s+6)^2/4 \rfloor\}$.*

We remark that $\lfloor (s+6)^2/4 \rfloor \ge 8s$ for $s \ge 18$. That the upper bound of $8s$ on the product $\gamma_t^1\gamma_t^2$ is achievable may be seen by taking $G_1 = sK_2$ or $G_2 = sK_2$. If $G_1 \cong B_{12}$, where B_{12} is the bipartite cubic graph of order $n = 12$ shown in Fig. 15.3, then $G_2 \cong B_{12}$ and $\gamma_t^1\gamma_t^2 = 36 = \lfloor (s+6)^2/4 \rfloor$. This shows that the product $\gamma_t^1\gamma_t^2$ may be equal to $\lfloor (s+6)^2/4 \rfloor$. However this example when $G_1 \cong B_{12}$ illustrating that $\gamma_t^1\gamma_t^2 = \lfloor (s+6)^2/4 \rfloor$ is possible has $s = 6$ and $8s > \lfloor (s+6)^2/4 \rfloor$. We close this section with the following question (see Question 18.15 in Sect. 18.14): Is it true that if $G_1 \oplus G_2 = K(s,s)$ where neither G_1 nor G_2 have an isolated vertex, then $\gamma_t(G_1)\gamma_t(G_2) \le 8s$?

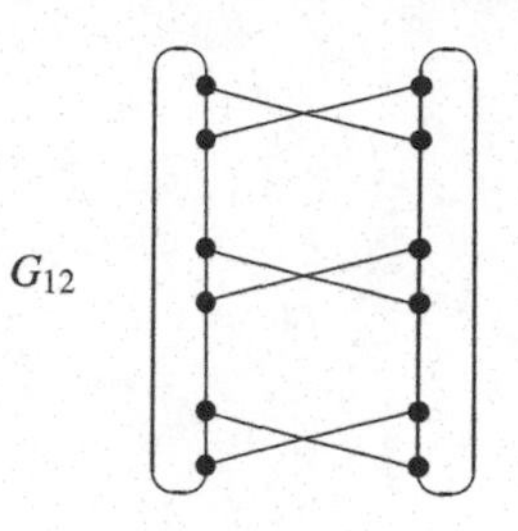

Fig. 15.3 The bipartite cubic graph B_{12}

15.4 Multiple Factor Nordhaus–Gaddum-Type Results

Recall that if $G_1, G_2, \ldots, G_k$ are graphs on the same vertex set but with pairwise disjoint edge sets, then $G_1 \oplus G_2 \oplus \cdots \oplus G_k$ denotes the graph whose edge set is the union of their edge sets.

Plesník [173] was the first to extend Nordhaus–Gaddum results to the case where the complete graph is factored into more than two factors. This approach was continued in [77] where the domination number and $G_1 \oplus G_2 \oplus G_3 = K_n$ are considered. In this section, we examine the sum and product of $\gamma_t(G_1), \gamma_t(G_2), \ldots, \gamma_t(G_k)$, where $G_1 \oplus G_2 \oplus \cdots \oplus G_k = K_n$. The following two results can be found in [112].

Theorem 15.14 ([112]). *If $G_1 \oplus G_2 \oplus \cdots \oplus G_k = K_n$, where $n \geq 7$, $3 \leq k \leq n-2$ and $\delta(G_i) \geq 1$ for each $i \in \{1,2,\ldots,k\}$, then*

$$\sum_{j=1}^{k} \gamma_t(G_j) \leq (k-1)(n+1).$$

For specific values of k the bound in Theorem 15.4 is shown to be sharp in [112]. Since the geometric mean of a set of positive numbers is less than or equal to their arithmetic mean, we have the following immediate consequence of Theorem 15.4.

Corollary 15.15 ([112]). *Let $G_1 \oplus G_2 \oplus \cdots \oplus G_k = K_n$, where $n \geq 7$, $3 \leq k \leq n-2$, and $\delta(G_i) \geq 1$ for each $i \in \{1,2,\ldots,k\}$. Then,*

$$\prod_{j=1}^{k} \gamma_t(G_j) \leq \left((1-\frac{1}{k})(n+1)\right)^k.$$

For $k \geq 3$, we do not know if the upper bound of Corollary 15.15 is tight. However it is unlikely that the upper bound is optimal. It is indeed a natural question to ask whether the upper bound of Corollary 15.15 can be improved by using a stronger argument than simply applying the fact that the geometric mean is less than or equal to the arithmetic mean. However we do not have any guess what the correct upper bound should be.

Chapter 16
Upper Total Domination

16.1 Upper Total Domination

In this chapter we focus on the upper total domination number of a graph. Recall that the *upper domination number* of a graph G, denoted by $\Gamma(G)$, is the maximum cardinality of a minimal dominating set in G, while the *upper total domination number* of G, denoted by $\Gamma_t(G)$, is the maximum cardinality of a minimal TD-set in G. We call a minimal dominating set of cardinality $\Gamma(G)$ a $\Gamma(G)$-set. Similarly, we call a minimal TD-set of cardinality $\Gamma_t(G)$ a $\Gamma_t(G)$-set.

16.1.1 Trees

The upper total domination number of a path is established in [52].

Proposition 16.1 ([52]). *For $n \geq 2$ an integer, $\Gamma_t(P_n) = 2\lfloor (n+1)/3 \rfloor$.*

Chellali, Favaron, Haynes, and Raber [28] determined upper bounds on the upper total domination number of a tree. Recall that the independence number $\alpha(G)$ of G is the maximum cardinality of an independent set of vertices of G, while the 2-independence number $\alpha_2(G)$ of G is the maximum cardinality of set of vertices that induce a subgraph of maximum degree at most 1 in G.

Theorem 16.2 ([28]). *Let T be a nontrivial tree. Then;*

(a) $\Gamma_t(T) \leq \alpha_2(T)$*, and this bound is sharp.*
(b) $\Gamma_t(T) \leq 2\gamma(T)$*, and this bound is sharp.*
(c) $\Gamma_t(T) \leq 2\alpha(T) - 1$*, and the bound of 2 on $\Gamma_t(T)/\alpha(T)$ is asymptotically sharp.*

M.A. Henning and A. Yeo, *Total Domination in Graphs*, Springer Monographs in Mathematics, DOI 10.1007/978-1-4614-6525-6_16,

16.1.2 Upper Total Domination Versus Upper Domination

The upper total domination number and upper domination number of a graph with no isolated vertex are incomparable. For example, if G is a complete bipartite graph $K_{k,k}$, then $\Gamma(G) = k$ and $\Gamma_t(G) = 2$, while if H is obtained from a star $K_{1,k}$ by subdividing every edge exactly once, then $\Gamma(H) = k+1$ and $\Gamma_t(H) = 2k$. Hence for every positive integer k, there exist graphs G and H such that $\Gamma(G) - \Gamma_t(G) = k$ and $\Gamma_t(H) - \Gamma(H) = k$.

However there is a relationship between the upper total domination number and upper domination number of a graph, as first observed by Dorbec, Henning, and Rall [53]. In order to state this relationship, we shall need the following lemma.

Lemma 16.3 ([53]). *If G is a graph with no isolated vertex, then every $\Gamma_t(G)$-set contains as a subset a minimal dominating set S such that $|S| \geq \frac{1}{2}\Gamma_t(G)$ and* $\mathrm{epn}(v,S) \geq 1$ *for each $v \in S$.*

Let D be a $\Gamma_t(G)$-set of a graph G with no isolated vertex. By Lemma 16.3, there is a minimal dominating set S of G such that $S \subseteq D$ and $|S| \geq \frac{1}{2}\Gamma_t(G)$, implying that $\Gamma(G) \geq |S| \geq \frac{1}{2}\Gamma_t(G)$. Furthermore, we observe that if G has order $n \geq 2$, then $\Gamma(G) \leq n-1$, with equality if and only if G is a star $K_{1,n-1}$. Therefore we have the following relationship first established in [53], where the lower bound, which is tight only for stars $K_{1,n-1}$, follows trivially from the fact that $\Gamma_t(G) \geq 2 \geq 2\Gamma(G)/(n-1)$.

Theorem 16.4 ([53]). *For any graph G of order n with no isolated vertex,*

$$\left(\frac{2}{n-1}\right)\Gamma(G) \leq \Gamma_t(G) \leq 2\Gamma(G).$$

Dorbec, Henning, and Rall [53] observed that a graph with upper total domination number equal to twice its upper domination number has the following structural properties.

Observation 16.5 ([53]). *If $G = (V,E)$ is a graph with no isolated vertex satisfying $\Gamma_t(G) = 2\Gamma(G)$, then the graph G has the following three properties:*

(a) *Every $\Gamma_t(G)$-set induces a subgraph that consists of disjoint copies of K_2.*
(b) *For every $\Gamma_t(G)$-set S, every vertex of $V \setminus S$ is contained in a common triangle with two vertices of S.*
(c) $\Gamma(G) = \alpha(G)$.

However it remains an open problem to find a characterization of graphs G satisfying $\Gamma_t(G) = 2\Gamma(G)$ (see Open Problem 18.1 in Sect. 18.10).

16.2 Bounds on the Upper Total Domination Number

In this section we discuss upper bounds on upper total domination number of a graph in terms of its order and its minimum degree. Even if the minimum degree is large, the upper total domination number can be made arbitrarily close to the order of the graph.

Theorem 16.6 ([62]). *If G is a connected graph of order $n \geq 3$, then $\Gamma_t(G) \leq n-1$. Furthermore, if G has minimum degree $\delta \geq 2$, then $\Gamma_t(G) \leq n-\delta+1$, and this bound is sharp.*

Proof. Let $G=(V,E)$ and let $v \in V$ be an arbitrary vertex, which is not a support vertex. Then, $V \setminus \{v\}$ is a TD-set in G, implying that the set V is not a minimal TD-set and so $\Gamma_t(G) \leq n-1$. That this bound is sharp may be seen by taking G to be the graph of order $n=2k+1$ obtained from a star $K_{1,k}$ by subdividing every edge exactly once. If v denotes the central vertex of the star, then $V \setminus \{v\}$ is a minimal TD-set on G, and so $\Gamma_t(G) \geq n-1$. Consequently, $\Gamma_t(G) = n-1$.

Suppose that $\delta \geq 2$. If $S \subseteq V$ is a set of size at least $n-\delta+1$ and v is an arbitrary vertex in V, then since $d_{V\setminus S}(v) \leq |V \setminus S| = n-|S| < \delta$, we have that $d_S(v) = d_G(v) - d_{V\setminus S}(v) > d_G(v) - \delta \geq 0$. Hence every vertex in G is adjacent to a vertex of S. Therefore, S is a TD-set in G, implying that no minimal TD-set has size greater than $n-\delta+1$. This proves that $\Gamma_t(G) \leq n-\delta+1$.

The complete graph $K_{\delta+1}$ shows that the bound is sharp as $n(K_{\delta+1}) - \delta + 1 = 2 = \Gamma_t(K_{\delta+1})$. □

For any fixed $\delta \geq 2$ there in fact exist infinitely many graphs, G, with minimum degree δ and with $\Gamma_t(G) = n(G) - \delta + 1$. Such a family of graphs can be constructed as follows. Let $x \geq \lceil \delta/2 \rceil$ be arbitrary. Let G_x be obtained from a complete bipartite graph $K_{2x,\delta-1}$ with partite sets X and Y where $|X| = 2x$ by adding a perfect matching between the vertices of X (and so, $G[X] = xK_2$). Then, G_x is a connected graph of order $n = 2x+\delta-1$ with minimum degree δ. Since X is a minimal TD-set of G_x, $\Gamma_t(G_x) \geq |X| = 2x = n-\delta+1$. Consequently, $\Gamma_t(G_x) = n-\delta+1$.

16.3 Upper Total Domination in Claw-Free Graphs

In this section, we show that the upper bound of Theorem 16.6 can be improved if we restrict our attention to claw-free graphs. For this purpose, we first define two families $\mathscr{F}$ and $\mathscr{G}$ of connected claw-free graphs constructed in [62].

For $\delta \geq 5$, let F be a connected graph with minimum degree at least δ whose vertex set can be partitioned into two sets V_1 and V_2 where the edges between V_1 and V_2 induce a perfect matching and where the subgraph induced by each of V_1 and V_2 is a clique or a disjoint union of cliques. Let $\mathscr{F}$ denote the family of all such graphs F. One can show that $\Gamma_t(G) = |V(G)|/2$, for all $G \in \mathscr{F}$ with $\delta(G) \geq 2$. This is not true when $\delta(G) = 1$ though as can be seen by considering $P_6 \in \mathscr{F}$.

Fig. 16.1 A graph $G \in \mathscr{G}_2$

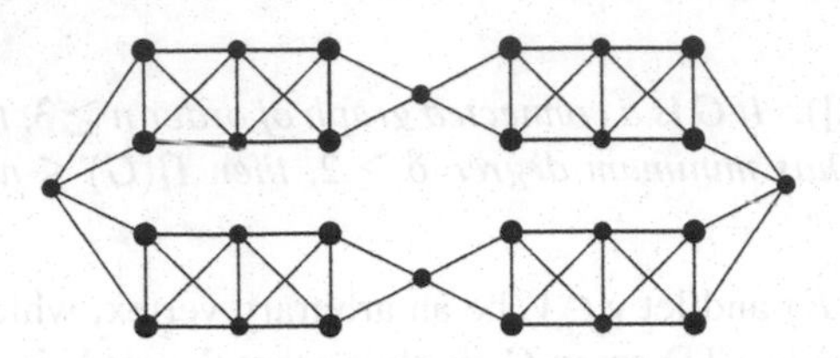

Fig. 16.2 A graph $G \in \mathscr{G}_4$

Let $\mathscr{G}$ be the family of connected claw-free graphs $G = (V, E)$ that admit a vertex partition $V = X \cup C$ such that $G[X] = qK_2$ and each vertex of C is adjacent to vertices of X from exactly two K_2s and possibly to other vertices of C. It is shown in [62] that if $G \in \mathscr{G}$, then $|X| \leq 2|C| + 2$. Further if G has minimum degree $\delta \geq 3$, then $|X| \leq 4(\delta - 1)/|C|$. For $\delta \in \{1, 2\}$, let $\mathscr{G}_\delta = \{G \in \mathscr{G} \colon \delta(G) \geq \delta \text{ and } |X| = 2|C| + 2\}$, while for $\delta \geq 3$, let $\mathscr{G}_\delta = \{G \in \mathscr{G} \colon \delta(G) \geq \delta \text{ and } |X| = 4|C|/(\delta - 1)\}$.

The path $P_{3k+2} \colon v_1, v_2, \ldots, v_{3k+2}$ on $3k+2$ vertices, where $k \geq 1$, is an example of a graph in the family $\mathscr{G}_1$. Let $C = \{v_{3i} \mid i = 1, \ldots, k\}$ and let $X = V(P_{3k+2}) \setminus C$. Then, $G[X] = (k+1)K_2$, and each vertex of C is adjacent to vertices of X from exactly two K_2s. Hence, $P_{3k+2} \in \mathscr{G}$. Since $|X| = 2|C| + 2$, $P_{3k+2} \in \mathscr{G}_1$.

Examples of graphs in the families $\mathscr{G}_2$ and $\mathscr{G}_4$, for example, are shown in Figs. 16.1 and 16.2, respectively, where the large vertices form the set X and the small vertices the set C. The graph obtained from the graph G of Fig. 16.1 by adding a new (small) vertex (to the set C) and joining it to the four vertices of degree 2 in G is an example of a graph in $\mathscr{G}_3$. We remark that each of the families $\mathscr{G}_i$ with $i \in \{1, 2, 3, 4, 5\}$ contains graphs of arbitrarily large order.

We are now in a position to state the result from [62] which provides tight upper bounds on the upper total domination number of a connected claw-free graph in terms of its order.

Theorem 16.7 ([62]). *If G is a connected claw-free graph of order n and minimum degree δ, then*

$$\Gamma_t(G) \leq \begin{cases} \dfrac{2(n+1)}{3} & \text{if } \delta \in \{1,2\} \\[2ex] \dfrac{4n}{\delta + 3} & \text{if } \delta \in \{3,4,5\} \\[2ex] \dfrac{n}{2} & \text{if } \delta \geq 6 \end{cases}$$

Furthermore, $\Gamma_t(G) = 2(n+1)/3$ *if and only if* $G \in \mathscr{G}_1 \cup \mathscr{G}_2$. *If* $\delta \in \{3,4,5\}$, *then* $\Gamma_t(G) = 4n/(\delta+3)$ *if and only if* $G \in \mathscr{G}_\delta$ *or* $\delta = 5$ *and* $G \in \mathscr{F}$. *If* $\delta \geq 6$, *then* $\Gamma_t(G) = n/2$ *if and only if* $G \in \mathscr{F}$.

We remark that the upper bounds in Theorem 16.7 are sharp even for connected claw-free graphs of arbitrarily large order.

16.4 Upper Total Domination in Regular Graphs

Recall that by Theorem 16.6 if G is a connected graph G of order $n \geq 3$, then $\Gamma_t(G) \leq n-1$, and this bound is sharp even for graphs of arbitrarily large order. However if we impose a regularity condition on the graph, then this bound can be greatly improved using edge weighting functions on total dominating sets.

We will outline how to obtain this improvement below. We adopt the notation from [191]. Let $G = (V,E)$ be a graph and let S be a TD-set in G. Now define the *edge weight function* $\psi_S : E \to [0,1]$ and *vertex weight function* $\phi_S : V \to [0,\infty]$ as follows:

(a1): If $e \in [S,S]$ or if $e \in [V \setminus S, V \setminus S]$, then $\psi_S(e) = 0$.
(a2): If $e \in [S, V \setminus S]$ joins a vertex $v \in V \setminus S$ to S, then $\psi_S(e) = 1/d_S(v)$.
(b): For all $v \in V$, let

$$\phi_S(v) = \sum_{u \in N(v)} \psi_S(uv).$$

That is, ϕ_S assigns to each vertex $v \in V$ the sum of the weights of the edges incident with v.

Since S is a TD-set in G, for each $v \in V \setminus S$, the sum of the weights of the edges incident with v, $\phi_S(v)$, is 1. Furthermore as all nonzero weighted edges join a vertex of S and a vertex $V \setminus S$, we get the following equation since each value in the equation below is equal to the sum of the weights of all edges in G:

$$\sum_{v \in S} \phi_S(v) = \sum_{e \in E} \psi_S(e) = \sum_{w \in V \setminus S} \phi_S(v) = |V \setminus S|. \tag{16.1}$$

We are now in position to state the following result which shows that if we impose a regularity condition on the graph, then the bound on its upper total domination number can be greatly improved.

Theorem 16.8 ([191]). *For every* k*-regular graph* G *of order* n *with no isolates,* $\Gamma_t(G) \leq n/(2 - \frac{1}{k})$.

Proof. Let G be a k-regular graph on n vertices where $k \geq 1$ and let S be a $\Gamma_t(G)$-set. We use the edge weight function ψ_S and vertex weight function ϕ_S to count the number of vertices in S relative to n. We show that, on average, $\phi_S(v) \geq 1 - \frac{1}{k}$ for each vertex $v \in S$.

Let C be any component of $G[S]$. Since S is a TD-set in G, we note that $|V(C)| \geq 2$. First we consider the case when $|V(C)| = 2$. By the k-regularity of G, each vertex in C has $k-1$ edges to $V \setminus S$, each of weight at least $1/k$, implying that $\phi_S(v) \geq (k-1)/k$ for each vertex v in C, as desired.

Next we consider the case when $|V(C)| \geq 3$. Let A be the set of vertices of degree 1 in C and let $B = N_C(A)$. We note that $|A| \geq |B|$ since there are $|A|$ edges in $[A,B]$ and every vertex of B is adjacent to at least one such edge. Since $|V(C)| \geq 3$, every vertex in B has degree at least 2 in C. Thus for every $v \in A$, we have that $\text{ipn}(v,S) = \emptyset$ which implies that $\text{epn}(v,S) \neq \emptyset$, and therefore some edge incident with the vertex v has weight 1. Since $d_{V \setminus S}(v) = k-1$ and since each such edge joining v to $V \setminus S$ has weight at least $1/k$, we therefore have that $\phi_S(v) \geq 1 + (k-2)/k$. Furthermore for each vertex $v \in V(C) \setminus (A \cup B)$, we have $\text{ipn}(v,S) = \emptyset$ which implies that $\text{epn}(v,S) \neq \emptyset$ and therefore that some edge incident with v has weight 1. Hence,

$$\begin{aligned}
\sum_{v \in V(C)} \phi_S(v) &\geq \sum_{v \in A} \phi_S(v) + \sum_{v \in V(C) \setminus (A \cup B)} \phi_S(v) \\
&\geq \left(\frac{2(k-1)}{k}\right)|A| + |V(C) \setminus (A \cup B)| \\
&\geq \left(\frac{k-1}{k}\right)(|A| + |B|) + \left(\frac{k-1}{k}\right)|V(C) \setminus (A \cup B)| \\
&= \left(\frac{k-1}{k}\right)|V(C)|.
\end{aligned}$$

Since C is an arbitrary component of $G[S]$, we have therefore shown that on average, $\phi_S(v) \geq 1 - \frac{1}{k}$ for each vertex $v \in S$. Hence by Theorem 16.1, we have that

$$n - |S| = |V \setminus S| = \sum_{v \in S} \phi_S(v) \geq \left(\frac{k-1}{k}\right)|S|.$$

Therefore, $n \geq (1 + \frac{k-1}{k})|S|$, and so $\Gamma_t(G) = |S| \leq n/(2 - \frac{1}{k})$. This establishes the desired upper bound. □

The extremal graphs that achieve equality in the upper bound of Theorem 16.8 are characterized in [191]. We note that in order to get equality in Theorem 16.8 we need the following to be true in the proof of Theorem 16.8. If $|V(C)| > 2$, then $|A| = |B|$ and $V(C) = A \cup B$. Further, all vertices $a \in A$ have $\phi_S(a) = 1 + (k-2)/k$ and all vertices $b \in B$ have $\phi_S(b) = 0$. If $|V(C)| = 2$, then both vertices of C have vertex weight $(k-1)/k$. Let $\mathscr{G}'$ be the family of all graphs which have a set S with the above properties. Given this information it is now possible to give an explicit description of the family $\mathscr{G}'$ and this is done in [191].

More precisely, the family $\mathscr{G}'$ is constructed as follows. A (k,s,t)-*triple* is defined in [191] as three nonnegative integers k, s, and t satisfying the following four conditions:

- $2s + t \geq k \geq 1$.
- $2(s+t) = \ell k$ for some positive integer ℓ.

- If $k = 1$, then $t = 0$.
- If $t > 0$, then $t \geq k$ where t is even whenever k is even.

Given a (k,s,t)-triple, a (k,s,t)-graph is defined in [191] as follows. If $k = 1$ a *(k,s,t)-graph* is defined to be the empty graph on $2s$ vertices. For $k \geq 2$, a *(k,s,t)-graph* is defined to be any bipartite graph, G, with partite sets $X = X_1 \cup X_2$ and Y such that $|X| = 2s+t$, $|X_1| = 2s$, $|X_2| = t$, and $|Y| = 2s+t-\ell$, and for all $x_1 \in X_1$, $x_2 \in X_2$, and $y \in Y$, we have $d_G(x_1) = k-1$, $d_G(x_2) = k-2$, and $d_G(y) = k$. We remark that (k,s,t)-graphs exist for every (k,s,t)-triple.

Let $\mathscr{G}'$ be the family of regular graphs (not necessarily connected) constructed in [191] as follows. Let G_1 be a (k,s,t)-graph. Let $U = \{u_1, v_1, u_2, v_2, \ldots, u_s, v_s\}$ be the set of $2s$ vertices in G_1 of degree $k-1$, let $W = \{w_1, w_2, \ldots, w_t\}$ be the set of t vertices in G_1 of degree $k-2$, and let $Z = \{z_1, z_2, \ldots, z_{2s+t-\ell}\}$ be the set of $2s+t-\ell$ vertices in G_1 of degree k. We remark that if $s = 0$, then $U = \emptyset$, and if $t = 0$, then $W = \emptyset$. If $s = 0$, let $E_U = \emptyset$, while if $s \geq 1$, let $E_U = \{u_1v_1, \ldots, u_sv_s\}$. If $t = 0$, let G be the k-regular graph obtained from G_1 by adding the edges from the set E_U. For $t \geq 1$, let G_2 and G_3 be disjoint $(k-1)$-regular graphs (not necessarily connected) of order t with $X = V(G_2) = \{x_1, x_2, \ldots, x_t\}$ and $Y = V(G_3) = \{y_1, y_2, \ldots, y_t\}$, and let $E_W = \bigcup_{i=1}^{t} \{x_iw_i, w_iy_i\}$. If $t \geq 1$, let G be the k-regular graph obtained from the disjoint union $G_1 \cup G_2 \cup G_3$ by adding the edges from the set $E_U \cup E_W$. Let $\mathscr{G}'$ be the family of all graphs G thus constructed. The construction of a k-regular graph G in the family $\mathscr{G}'$ is illustrated in Fig. 16.3. When $k = 3$, for example, two graphs in the family $\mathscr{G}'$ are shown in Fig. 16.4, where the one graph is constructed from a (3, 3, 0)-graph and the other graph from a (3, 0, 3)-graph and two copies of $\mathscr{K}$.

We are now in a position to characterize the extremal regular graphs with large upper total domination number.

Theorem 16.9 ([191]). *If G is a k-regular graph of order n with no isolate vertices satisfying $\Gamma_t(G) = n/(2-\frac{1}{k})$, then $G \in \mathscr{G}'$.*

16.5 A Vizing-Like Bound for Upper Total Domination

In this section, we consider Vizing-like bounds for the upper total domination number of Cartesian products of graphs. We begin with the following trivial lemma.

Lemma 16.10 ([53]). *If $H = K_2$ and G is any graph with no isolated vertex, then*

$$\Gamma_t(G)\Gamma_t(H) \leq 2\Gamma_t(G \Box H),$$

with equality if and only if G is a disjoint union of copies of K_2.

Proof. Let $V(H) = \{u, v\}$. Then, $V(G) \times \{v\}$ is a minimal TD-set of $G \Box H$, and so $\Gamma_t(G \Box H) \geq |V(G)| \geq \Gamma_t(G) = \frac{1}{2}\Gamma_t(G)\Gamma_t(H)$. Further, if $\Gamma_t(G \Box H) = \frac{1}{2}\Gamma_t(G)\Gamma_t(H)$, then we must have equality throughout this inequality chain. In particular, $\Gamma_t(G) = |V(G)|$, implying that G is a disjoint union of copies of K_2. Conversely if $G = kK_2$

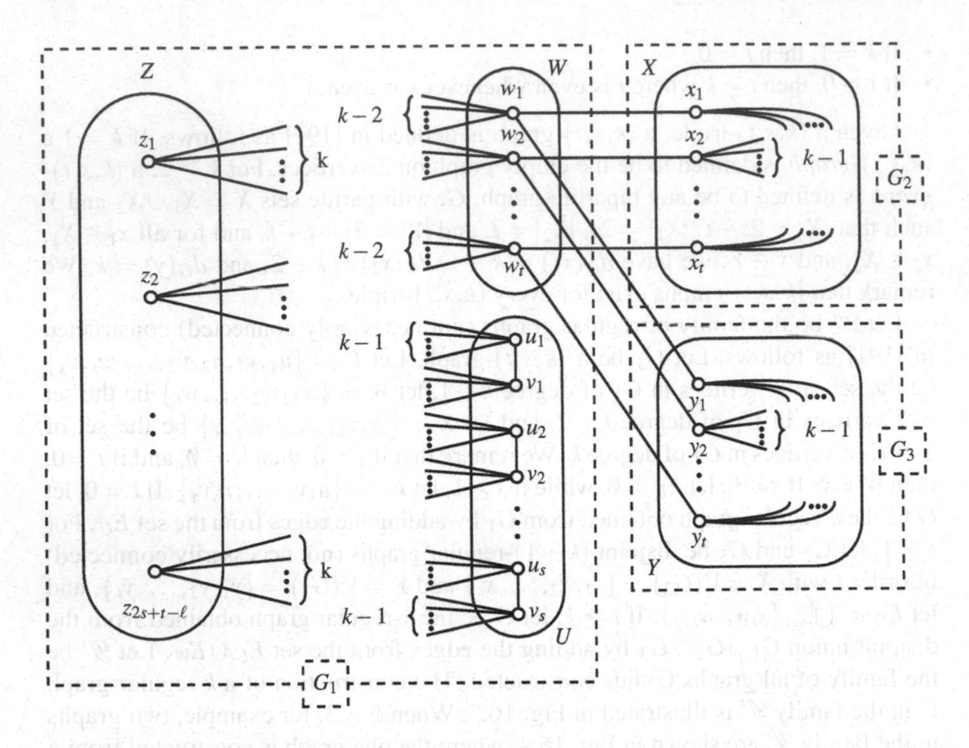

Fig. 16.3 The construction of a k-regular graph G in the family $\mathscr{G}'$ from the bipartite (k,s,t)-graph G_1 with partite sets Z and $U \cup W$ and the $(k-1)$-regular graphs G_2 and G_3

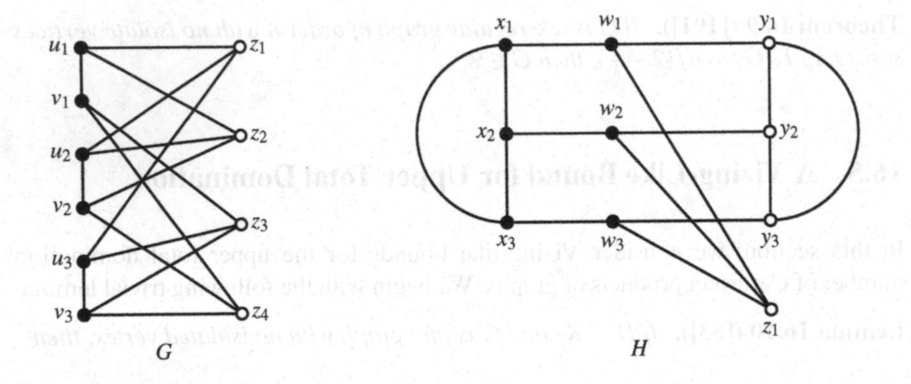

Fig. 16.4 Two graphs from the family $\mathscr{G}_1$: the graph G, constructed from a $(3,3,0)$-graph, and the graph H, constructed from a $(3,0,3)$-graph and two copies of K_3. In each case, the darkened vertices depict a minimal TD-set of maximum cardinality

for some $k \geq 1$, then $G \square H = kC_4$ and $2\Gamma_t(G \square H) = 4k = (2k)2 = \Gamma_t(G)\Gamma_t(H)$. Hence if G is a disjoint union of copies of K_2, then $\Gamma_t(G)\Gamma_t(H) = 2\Gamma_t(G \square H)$. □

Using Lemma 16.3, the following Vizing-like bound for the upper total domination number of Cartesian products of graphs is established in [53]. Recall the notion of a fiber of a graph introduced in Sect. 12.2. For a vertex g of G, *H-fiber*, ${}^gH$, is the subgraph of $G \square H$ induced by the set $\{(g,h) \mid h \in V(H)\}$. Similarly, for $h \in H$, the *G-fiber*, G^h, is the subgraph induced by $\{(g,h) \mid g \in V(G)\}$. We note that all G-fibers are isomorphic to G and all H-fibers are isomorphic to H.

Theorem 16.11 ([53]). *If G and H are connected graphs of order at least 2, then $\Gamma_t(G)|V(H)| \leq 2\Gamma_t(G \square H)$ and $\Gamma_t(H)|V(G)| \leq 2\Gamma_t(G \square H)$.*

Proof. If $G = K_2$ or $H = K_2$, then the result follows from Lemma 16.10. Hence we may assume that both G and H have order at least 3. By Lemma 16.3, there exists a minimal dominating set S of G such that $|S| \geq \frac{1}{2}\Gamma_t(G)$ and $\text{epn}(v,S) \geq 1$ for each $v \in S$. We now consider the set $D = S \times V(H)$. Since S dominates $V(G)$, the set D dominates $G \square H$. Further, for every vertex $g \in S$, the vertices in $\{g\} \times V(H)$ are totally dominated by their neighbors in the H-fiber ${}^gH$, implying that D is a TD-set of $G \square H$. We show that D is in fact a minimal TD-set of $G \square H$. Let $v \in D$. Then, $v = (g,h)$ for some vertices $g \in S$ and $h \in V(H)$. Let $v' = (g',h)$ where $g' \in \text{epn}(g,S)$ in G. Then, $v' \in \text{epn}(v,D)$ in $G \square H$. Hence in $G \square H$, $|\text{epn}(v,D)| \geq 1$ for every $v \in D$. Thus, by Proposition 2.1, D is a minimal TD-set of $G \square H$, and so $\Gamma_t(G \square H) \geq |D|$. Consequently,

$$\Gamma_t(G \square H) \geq |D| = |S| \times |V(H)| \geq \frac{1}{2}\Gamma_t(G)|V(H)|,$$

or, equivalently, $\Gamma_t(G)|V(H)| \leq 2\Gamma_t(G \square H)$. Analogously, we have that $\Gamma_t(H)|V(G)| \leq 2\Gamma_t(G \square H))$. □

A connected graph that can be constructed from $k \geq 2$ disjoint copies of K_3 by identifying a set of k vertices, one from each K_3, into one vertex is called a *daisy* with *k petals*. It is proven in [53] that if G and H are both daisies with k petals, then $\Gamma_t(G)|V(H)| = \Gamma_t(H)|V(G)| = 2\Gamma_t(G \square H)$. Hence the bound in Theorem 16.11 is sharp. However it remains an open problem to characterize the graphs G and H achieving equality in the bound of Theorem 16.11.

As an immediate consequence of Lemma 16.10 and Theorem 16.11, we have the following result.

Theorem 16.12 ([53]). *Let G and H be graphs with no isolated vertices. Then, $\Gamma_t(G)\Gamma_t(H) \leq 2\Gamma_t(G \square H)$, with equality if and only if both G and H are a disjoint union of copies of K_2.*

Chapter 17
Variations of Total Domination

17.1 Introduction

There are several variations of total dominating sets in graphs. In this chapter, we select four such variations of a TD-set in a graph and briefly discuss each variation. In all cases, the TD-set is required to satisfy further properties in addition to totally dominating the vertex set of the graph.

17.2 Total Restrained Domination

Among all the variations of TD-sets in graphs, perhaps the one most studied is that of a total restrained dominating set. A TD-set S in a graph $G = (V, E)$ with the additional property that every vertex not in S is adjacent to a vertex in $V \setminus S$ is called a *total restrained dominating set*, abbreviated TRD-set, in G. The minimum cardinality of a TRD-set of G is the *total restrained domination number* of G, denoted by $\gamma_{tr}(G)$. The concept of total restrained domination in graphs was introduced by Telle and Proskurowski in [196], albeit indirectly, as a vertex partitioning problem. There are currently several dozen papers on this concept of total restrained dominating sets in graphs, including among others, [32, 81–83, 120, 146–149].

We mention here a small sample of selected results. We begin with an upper bound on the total restrained domination number in terms of the order and maximum degree of the graph.

Theorem 17.1 ([120]). *If G is a connected graph of order $n \geq 4$, maximum degree Δ where $\Delta \leq n-2$, and minimum degree at least 2, then*

M.A. Henning and A. Yeo, *Total Domination in Graphs*, Springer Monographs in Mathematics, DOI 10.1007/978-1-4614-6525-6_17,

$$\gamma_{tr}(G) \leq n - \frac{\Delta}{2} - 1,$$

and this bound is sharp.

If we restrict our attention to bipartite graphs, then we show that the bound of Theorem 17.1 can be improved.

Theorem 17.2 ([120]). *If G is a connected bipartite graph of order $n \geq 5$, maximum degree Δ where $3 \leq \Delta \leq n-2$, and minimum degree at least 2, then*

$$\gamma_{tr}(G) \leq n - \frac{2}{3}\Delta - \frac{2}{9}\sqrt{3\Delta - 8} - \frac{7}{9},$$

and this bound is sharp.

Joubert [149] studied the total restrained domination number in claw-free connected graphs. We define a *bow tie* as the graph on five vertices obtained from two disjoint triangles by identifying a vertex from each triangle into one new vertex.

Theorem 17.3 ([149]). *Let G be a connected claw-free graph of order n with $\delta(G) \geq 2$. If G is not a bow tie and if $G \notin \{C_3, C_5, C_6, C_7, C_{10}, C_{11}, C_{15}, C_{19}\}$, then $\gamma_{tr}(G) \leq 4n/7$.*

Using intricate and clever counting arguments, Jiang, Kang, and Shan [147] established the following upper bound on the total restrained domination number of a cubic graph.

Theorem 17.4 ([147]). *If G is a connected cubic graph of order n, then $\gamma_{tr}(G) \leq 13n/19$.*

The Jiang–Kang–Shan upper bound on $\gamma_{tr}(G)$ established in Theorem 17.4 was subsequently improved from $13n/19$ to $(n+4)/2$ in [190].

Theorem 17.5 ([190]). *If G is a connected cubic graph of order n, then $\gamma_{tr}(G) \leq (n+4)/2$.*

The following observation shows that the new improved bound on the total restrained domination number of a connected cubic graph established in Theorem 17.5 is essentially best possible.

Observation 17.6 ([190]). *If $G \in \mathcal{G} \cup \mathcal{H}$, where $\mathcal{G}$ and $\mathcal{H}$ are the two infinite families of graphs constructed in Theorem 5.10 or G is the generalized Petersen graph G_{16} shown in Fig. 1.2 and G has order n, then $\gamma_{tr}(G) = n/2$.*

However we conjecture that the constant term in the upper bound of Theorem 17.5 can be omitted. That is, we conjecture that if G is a connected cubic graph of order n, then $\gamma_{tr}(G) \leq n/2$. We state this conjecture as Conjecture 18.16 in Sect. 18.15.

If G is a 4-regular graph on n vertices, then by Theorem 13.15(a), there exist two disjoint TD-sets T_1 and T_2. Since both T_1 and T_2 are TRD-sets in G, we therefore have that $\gamma_{tr}(G) \leq \min\{|T_1|,|T_2|\} \leq n/2$. Hence as a consequence of Theorem 13.15(a), Conjecture 18.16 is true if the graph is 4-regular.

Theorem 17.7 ([141]). *If G is a 4-regular graph of order n, then $\gamma_{tr}(G) \leq n/2$.*

One can in fact say more for general $k \geq 4$. For this purpose, we define a set X of vertices in a hypergraph H to be a *free set* in H if we can 2-color $V(H) \setminus X$ such that every edge in H receives at least one vertex of each color. Equivalently, X is a free set in H if it is the complement of two disjoint transversals in H. The following results about free sets in hypergraph $H \in \mathscr{H}_k$ for k sufficiently large are established in [141].

Theorem 17.8 ([141]). *For every $k \geq 13$, every hypergraph $H \in \mathscr{H}_k$ of order n has a free set of size at least $n/14$.*

Theorem 17.9 ([141]). *Let $0 < \varepsilon < 1$ be arbitrary. For sufficiently large k, every hypergraph $H \in \mathscr{H}_k$ of order n has a free set of size at least $c_k n$, where*

$$c_k = 1 - 6(1+\varepsilon)\frac{\ln(k)}{k}.$$

Theorem 17.10 ([141]). *Let $0 < \varepsilon < 1$ be arbitrary. For sufficiently large k, there exist hypergraphs $H \in \mathscr{H}_k$ of order n such that every free set in H has size less than $c_k n$, where*

$$c_k = 1 - 2(1-\varepsilon)\frac{\ln(k)}{k}.$$

Let G be a k-regular graph on n vertices and let H be the ONH of G. Then, $H \in \mathscr{H}_k$ is a hypergraph on n vertices. Let X be a free set in H. Then we can partition $V(H) \setminus X$ into two sets T_1 and T_2 both of which are transversals in H. Hence both T_1 and T_2 are TRD-sets in G. Therefore, $\gamma_{tr}(G) \leq \min\{|T_1|,|T_2|\} \leq (n-|X|)/2$. Hence as a consequence of Theorems 17.8 and 17.9, we have the following improvements on the bound in Theorem 17.7 for sufficiently large k.

Theorem 17.11 ([141]). *For $k \geq 13$ if G is a k-regular graph of order n, then $\gamma_{tr}(G) \leq 13n/28$.*

Theorem 17.12 ([141]). *Let $0 < \varepsilon < 1$ be arbitrary. For sufficiently large k, if G is a k-regular graph of order n, then*

$$\gamma_{tr}(G) \leq 3(1+\varepsilon)\frac{\ln(k)}{k}.$$

We remark that if G is a k-regular graph, then Theorem 17.12 implies that as $k \to \infty$ we have $\gamma_{tr}(G)/|V(G)| \to 0$. This result significantly improves the best-known ratio which can be deduced from Theorem 17.7, namely, that $\gamma_{tr}(G)/|V(G)| \leq 1/2$ for every k-regular graph G with $k \geq 4$.

17.3 Double Total Domination

A TD-set S in a graph G is called a *double total dominating set*, abbreviated DTD-set, of G if every vertex of G is adjacent to at least two vertices in S. The minimum cardinality of a DTD-set in G is the double total domination number $\gamma_{\times 2,t}(G)$ of G. This parameter is studied, for example, in [135]. We remark that a DTD-set is also called a 2-*tuple total dominating set* in the literature. The more general concept of a *k-tuple total dominating set* S, where every vertex has at least k neighbors in S, is studied in [114] and elsewhere.

We mention here some selected results on double total domination in graphs. Recall that the Heawood graph is the graph shown in Fig. 5.11 (which is the unique 6-cage). The following results are established in [135].

Theorem 17.13 ([135]). *If G is a connected graph of order n with $\delta(G) \geq 3$, then $\gamma_{\times 2,t}(G) \leq 6n/7$ with equality if and only if G is the Heawood graph.*

Theorem 17.14 ([135]). *If G is a connected cubic graph of order n and G is not the Heawood graph, then $\gamma_{\times 2,t}(G) \leq 5n/6$ and this bound is sharp.*

That the bound of Theorem 17.14 is achievable may be seen by considering the cubic graph $G = G_{12}$ of order $n = 12$ shown in Fig. 17.1 that satisfies $\gamma_{\times 2,t}(G) = 10 = 5n/6$, due to the following. Since every pair of vertices in $X = \{x_1, x_2, \ldots, x_6\}$ have a common neighbor and G is cubic, every DTD-set in G must contain at least five vertices from this set. Analogously it must also contain at least five vertices from $Y = \{y_1, y_2, \ldots, y_6\}$, which implies that $\gamma_{\times 2,t}(G) \geq 10 = 5n/6$. Equality holds as we can take any five vertices from X and any five vertices from Y to obtain a DTD-set of size ten.

Theorem 17.15 ([135]). *If G is a connected graph of order n with $\delta(G) \geq 3$ and G is not the Heawood graph, then $\gamma_{\times 2,t}(G) \leq 11n/13$.*

We remark that the proofs of the above three results use hypergraph results, and these proof techniques demonstrate an interplay between strong transversals in hypergraphs and double total domination in graphs.

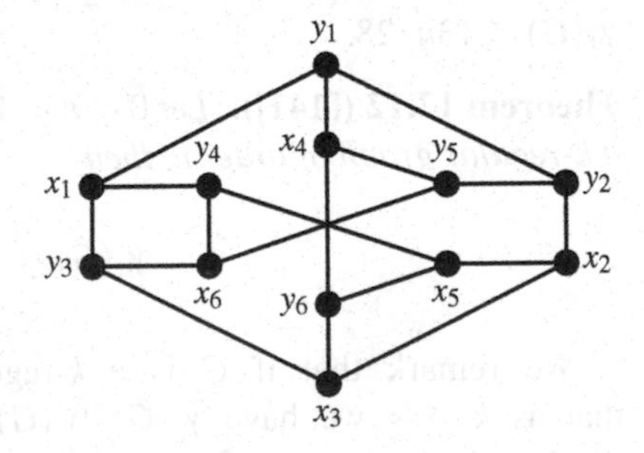

Fig. 17.1 The graph G_{12}

17.4 Locating Total Domination

The study of locating–dominating sets in graphs was pioneered by Slater [186, 187], and this concept was later extended to total domination in graphs. A *locating–total dominating set*, abbreviated LTD-set, in $G = (V,E)$ is a TD-set S with the property that distinct vertices in $V \setminus S$ are totally dominated by distinct subsets of S. Every graph G with no isolated vertex has a LTD-set since V is such a set. The *locating–total domination number*, denoted $\gamma_t^L(G)$, of G is the minimum cardinality of a LTD-set of G. This concept of locating–total domination in graphs was first studied by Haynes et al. [93] and has been studied, for example, in [13, 14, 27, 31, 33, 117, 123], and elsewhere.

We mention here some selected results on LTD-sets in cubic graphs. Let $\mathscr{G}_n$ denote the family of all connected cubic graphs of order n and let $\xi(n)$ be defined by

$$\xi(n) = \max\left\{\frac{\gamma_t^L(G)}{\gamma_t(G)}\right\},$$

where the maximum is taken over all graphs $G \in \mathscr{G}_n$.

Theorem 17.16 ([123]). *For $n \equiv 0 \pmod{16}$, we have $\xi(n) \geq \frac{4}{3}$. Further, there is an infinite family of connected claw-free cubic graphs of order n satisfying $\gamma_t^L(G)/\gamma_t(G) \geq 4/3$.*

Recall that by Theorem 5.7, if G is a connected cubic graph of order $n \geq 8$, then $\gamma_t(G) \leq n/2$. The extremal graphs (see Theorem 5.10) show that there are two infinite families, namely, $\mathscr{G}$ and $\mathscr{H}$, and one finite graph, namely, the generalized Petersen graph G_{16} of order 16 shown in Fig. 1.2, that achieve equality in this bound. Although these extremal graphs have equal total domination number and locating–total domination number, in general the locating–total domination number of a cubic graph can be very much larger than its total domination number as shown by Theorem 17.16. However, it is conjectured in [117] that except for the graphs K_4 and $K_{3,3}$, the tight upper bound on the total domination number of a connected cubic graph of one-half its order is surprisingly also an upper bound on its locating–total domination number. We state this conjecture as Conjecture 18.18 in Sect. 18.17.

For $k \geq 2$ an integer, let N_k be the connected cubic graph constructed as follows. Take k disjoint copies $D_1, D_2, \ldots, D_k$ of a diamond, where $V(D_i) = \{a_i, b_i, c_i, d_i\}$ and where a_ib_i is the missing edge in D_i. Let N_k be obtained from the disjoint union of these k diamonds by adding the edges $\{a_ib_{i+1} \mid i = 1, 2, \ldots, k-1\}$ and adding the edge a_kb_1. We call N_k a *diamond-necklace with k diamonds.* Let $\mathscr{N}_{\text{cubic}} = \{N_k \mid k \geq 2\}$. A diamond-necklace, N_8, with eight diamonds is illustrated in Fig. 17.2.

Let F_1, F_2, F_3, F_4, F_5 be the five graphs shown in Fig. 17.3 and let $\mathscr{F}_{\text{cubic}} = \{F_1, F_2, F_3, F_4, F_5\}$.

The following result confirms that Conjecture 18.18 is true for claw-free graphs.

Theorem 17.17 ([117]). *If $G \neq K_4$ is a connected cubic claw-free graph of order n, then $\gamma_t^L(G) \leq n/2$, with equality if and only if $G \in \mathscr{N}_{\text{cubic}} \cup \mathscr{F}_{\text{cubic}}$.*

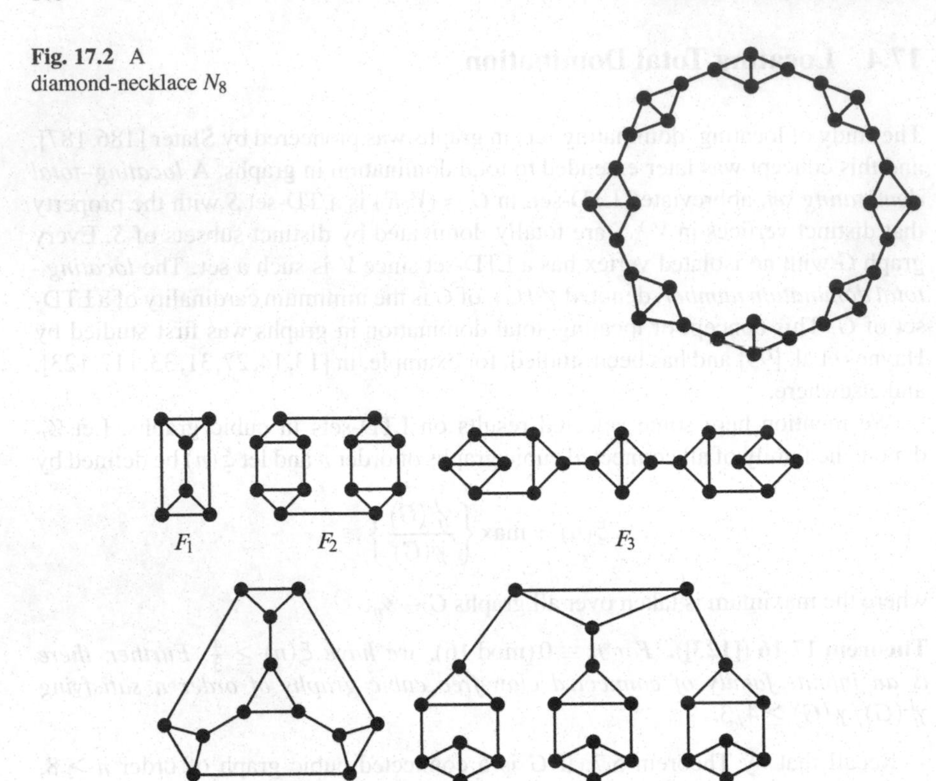

Fig. 17.2 A diamond-necklace N_8

Fig. 17.3 The graphs of the family $\mathscr{F}_{\text{cubic}}$

We remark that the structural requirement of claw-freeness is of interest since the locating–total domination number of a claw-free cubic graph tends to be very much larger than its total domination number. One might therefore expect that if Conjecture 18.18 is false, then it fails for the class of claw-free graphs. However Theorem 17.17 confirms that Conjecture 18.18 is true for claw-free graphs.

17.5 Differentiating Total Domination

A *differentiating total dominating set*, abbreviated differentiating TD-set, is similar to a LTD-set, except that in this case we impose the stricter requirement that distinct vertices, even vertices that belong to the TD-set, are totally dominated by distinct subsets of the TD-set. Hence a set S is a differentiating TD-set in a graph G if S is a TD-set in G with the additional property that $N(u) \cap S \neq N(v) \cap S$ for all distinct vertices u and v in G. We remark that a differentiating TD-set is also called *an identifying open code* or an *open locating–dominating set* or a *strong identifying code* in the literature. We shall adopt the terminology of an *identifying open code*, abbreviated IO-code. A *separating open code* in G is a set C of vertices such that $N(u) \cap C \neq N(v) \cap C \neq \emptyset$ for all distinct vertices u and v in G. Thus an IO-code is a set C that is both a separating open code and a TD-set. A graph is *twin-free* (or *open identifiable*) if every two distinct vertices has distinct open neighborhoods. A graph with no isolated vertex has an IO-code if and only if it is twin-free. If G is twin-free and G has no isolated vertex, we denote by $\gamma^{\rm IOC}(G)$ the minimum cardinality of an IO-code in G. The problem of identifying open codes was introduced by Honkala et al. [144] in the context of coding theory for binary hypercubes and further studied, for example, in [27, 140, 182, 183], and elsewhere.

A hypergraph H is *identifiable* if every two edges in H is distinct. Let $H = (V,E)$ be an identifiable hypergraph. We define a subset T of vertices in H to be a *distinguishing-transversal* if T is a transversal in H that distinguishes the edges, that is, $e \cap T \neq f \cap T$ for every two distinct edges e and f in H. A hypergraph has a distinguishing-transversal if and only if all its edges are distinct. The *distinguishing-transversal number* $\tau_D(H)$ of H is the minimum size of a distinguishing-transversal in H.

We note that if a graph G is twin-free with no isolated vertex, then the edges in its open neighborhood hypergraph, ONH(G), are distinct. Further a set is a distinguishing-transversal in ONH(G) if and only if it is an IO-code in G. Hence we have the following observation.

Observation 17.18 ([140]). *If G is a twin-free graph and $\delta(G) \geq 1$, then $\gamma^{\rm IOC}(G) = \tau_D({\rm ONH}(G))$.*

By Observation 17.18, identifying open codes in graphs can be translated to the problem of finding distinguishing-transversal in hypergraphs. The main advantage of considering hypergraphs rather than graphs is that the structure is easier to handle. The following hypergraph result is proven in [140].

Theorem 17.19 ([140]). *If H is a 3-uniform identifiable hypergraph of order n, size m, with $\Delta(H) \leq 3$, then $20\tau_D(H) \leq 12n + 3m$.*

If G is a twin-free cubic graph of order n, then the open neighborhood hypergraph, ONH(G), of G is a 3-regular, 3-uniform identifiable hypergraph of order n and size n. Hence as an immediate consequence of Observation 17.18 and Theorem 17.19, we have the following upper bound on the minimum cardinality of an IO-code.

Theorem 17.20 ([140]). *If G is a connected twin-free cubic graph of order n, then $\gamma^{\rm IOC}(G) \leq 3n/4$.*

The upper bound in Theorem 17.20 is achieved, for example, by the complete graph K_4 and the hypercube Q_3 shown in Fig. 17.4 that satisfy $\gamma^{\rm IOC}(K_4) = 3$ and $\gamma^{\rm IOC}(Q_3) = 6$.

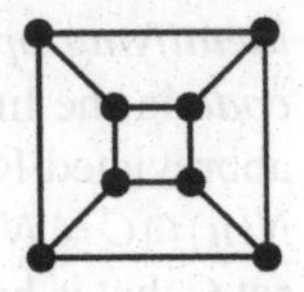
Fig. 17.4 The hypercube Q_3

However we conjecture that if G is a twin-free connected cubic graph of sufficiently large-order n, then $\gamma^{\mathrm{IOC}}(G) \le 3n/5$. We state this conjecture as Conjecture 18.19 in Sect. 18.18.

Chapter 18
Conjectures and Open Problems

18.1 Introduction

In this chapter, we list several conjectures and open problems which have yet to be settled or solved.

18.2 Total Domination Edge-Critical Graphs

By Theorem 11.6 in Chap. 11, we have that for every $k \geq 2$ there exists a k_tEC graph with diameter at least $\lfloor 3(k-1)/2 \rfloor$. So if the following conjecture is true, then the bound is tight.

Conjecture 18.1 ([127]). The maximum diameter of a k_tEC graph is $\lfloor 3(k-1)/2 \rfloor$ for every $k \geq 3$.

Furthermore, by Theorem 11.4, the above conjecture holds when $k \leq 6$.

18.3 Planar Graphs

By Theorem 6.5 in Chap. 11, if G is a planar graph of diameter 3, then $\gamma_t(G) \leq 10$, while if G has sufficiently large order, then $\gamma_t(G) \leq 7$. The following conjecture is posed in [107].

Conjecture 18.2. Every planar graph of diameter 3 has total domination number at most 6.

If Conjecture 18.2 is true, then the bound is sharp as shown by the graph of Fig. 18.1, which can be made arbitrarily large by duplicating any of the vertices of degree 2. This graph first appeared in the paper by MacGillivray and Seyffarth [164].

M.A. Henning and A. Yeo, *Total Domination in Graphs*, Springer Monographs in Mathematics, DOI 10.1007/978-1-4614-6525-6_18,

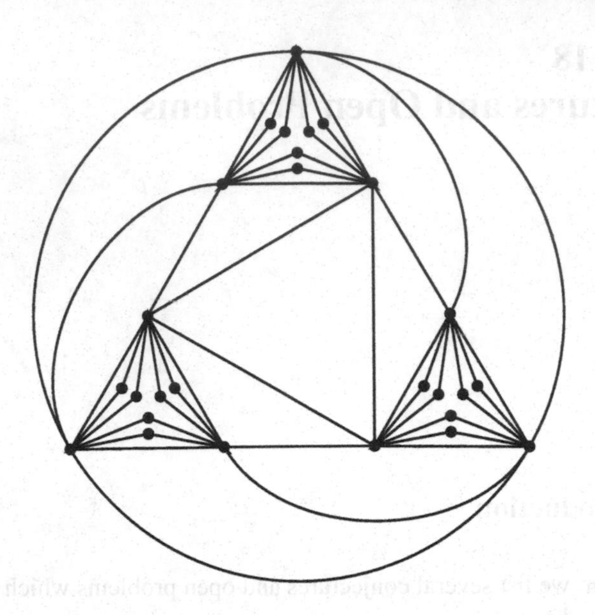

Fig. 18.1 A planar graph with diameter 3 and domination number 6

Furthermore, by adding edges joining vertices of degree 2, it is possible to construct such a planar graph with minimum degree equal to 3.

18.4 C_4-Free Graphs

In Theorem 5.7 it is shown that the total domination number of a connected graph with minimum degree three is never more than half the order of the graph. We note that each vertex in every graph of large order that achieves equality in the bound of Theorem 5.7 belongs to a 4-cycle. It is therefore a natural question to ask whether this upper bound of $n/2$ can be improved if we restrict G to contain no 4-cycles. We pose the following conjecture.

Conjecture 18.3. Let G be connected graph of order n with $\delta(G) \geq 3$ that contains no 4-cycles. If G is not the generalized Petersen graph G_{16} of order 16 shown in Fig. 1.2, then $\gamma_t(G) \leq 8n/17$.

Conjecture 18.3 claims that the absence of 4-cycles guarantees that the upper bound of $n/2$ for $\gamma_t(G)$ in Theorem 5.7 can be lowered to $8n/17$. If Conjecture 18.3 is true, then the bound is sharp as may be seen by considering the following family $\mathscr{L}$ of all graphs G that can be obtained from a connected C_4-free graph F with minimum degree at least 2 as follows: For each vertex v of F, add a copy G_v of the generalized Petersen graph, G_{16}, shown in Fig. 1.2 and join v to one vertex of G_v.

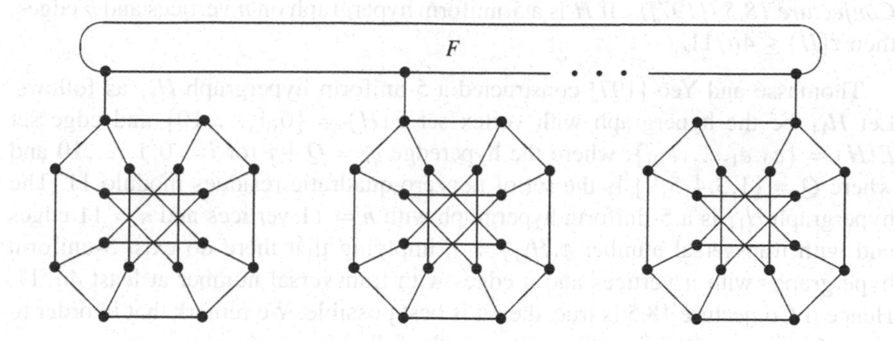

Fig. 18.2 A graph G in the family $\mathscr{L}$

A graph G in the family $\mathscr{L}$ is illustrated in Fig. 18.2. Each graph G of order n in the family $\mathscr{L}$ is a connected graph with no 4-cycles satisfying $\gamma_t(G) = 8n/17$.

18.5 Minimum Degree Four

By Theorem 5.18 in Sect. 5.6, if G is a connected graph of order n with $\delta(G) \geq 4$, then $\gamma_t(G) \leq 3n/7$, with equality if and only if G is the bipartite complement of the Heawood graph. We pose the following conjecture.

Conjecture 18.4. If G is connected graph of order n with $\delta(G) \geq 4$ that is not the bipartite complement of the Heawood graph, then $\gamma_t(G) \leq 2n/5$.

If Conjecture 18.4 is true, then the bound is sharp as may be seen by considering the following family $\mathscr{H}$ of all graphs that can be obtained from a connected graph F with minimum degree at least 3 as follows: For each vertex v of F, add a copy G_v of the bipartite complement of the Heawood graph and join v to one vertex of G_v. Each graph G of order n in the family $\mathscr{H}$ is a connected graph with $\delta(G) \geq 4$ satisfying $\gamma_t(G) = 2n/5$.

18.6 Minimum Degree Five

By Theorem 5.20 in Sect. 5.7, if H is a 5-uniform hypergraph on n vertices and m edges, then $\tau(H) \leq (10n + 7m)/44$, and by Theorem 5.21, if G is a connected graph of order n with $\delta(G) \geq 5$, then $\gamma_t(G) \leq 17n/44$. As remarked in Sect. 5.7, it is unlikely that the bounds of Theorems 5.20 and 5.21 are best possible. Thomassé and Yeo [197] made the following conjecture.

Conjecture 18.5 ([197]). If H is a 5-uniform hypergraph on n vertices and n edges, then $\tau(H) \leq 4n/11$.

Thomassé and Yeo [197] constructed a 5-uniform hypergraph H_{11} as follows. Let H_{11} be the hypergraph with vertex set $V(H) = \{0,1,\ldots,10\}$ and edge set $E(H) = \{e_0, e_1, \ldots, e_{10}\}$, where the hyperedge $e_i = Q + i$ for $i = 0,1,\ldots,10$ and where $Q = \{1,3,4,5,9\}$ is the set of nonzero quadratic residues modulo 11. The hypergraph H_{11} is a 5-uniform hypergraph with $n = 11$ vertices and $n = 11$ edges and with transversal number $\tau(H_{11}) = 4$, implying that there do exist 5-uniform hypergraphs with n vertices and n edges with transversal number at least $4n/11$. Hence if Conjecture 18.5 is true, then it is best possible. We remark that in order to prove Conjecture 18.5, it suffices to prove the following conjecture.

Conjecture 18.6. If H is a 5-uniform hypergraph on n vertices and m edges, then $\tau(H) \leq (9n+7m)/44$.

We also remark that Conjecture 18.5 would prove the following related conjecture also due to Thomassé and Yeo [197].

Conjecture 18.7 ([197]). If G is a connected graph of order n with $\delta(G) \geq 5$, then $\gamma_t(G) \leq 4n/11$.

18.7 Relating the Size and Total Domination Number

The following conjecture is posed in [43].

Conjecture 18.8 ([43]). If G is a bipartite graph without isolated vertices of order n, size m, and total domination number $\gamma_t \geq 3$ odd, then

$$m \leq \left\lfloor \frac{1}{4}((n-\gamma_t)(n-\gamma_t+4)+2\gamma_t-2) \right\rfloor.$$

Furthermore, the extremal graph is $G(n-k+3) \cup \left(\frac{k-3}{2}\right) K_2$, where $G(x)$ denotes the complete bipartite graph $K_{\lceil \frac{x}{2}-1 \rceil, \lfloor \frac{x}{2}+1 \rfloor}$ with a maximum matching removed.

18.8 Augmenting a Graph

Recall that Theorem 13.4 in Chap. 13 shows that if G is a graph on $n \geq 4$ vertices with minimum degree at least 2, then $\text{td}(G) \leq \frac{1}{4}(n - 2\sqrt{n}) + c\log n$ for some constant c. The following conjecture is posed in [54].

Conjecture 18.9. If G is a graph on $n \geq 4$ vertices with minimum degree at least 2, then $\text{td}(G) \leq \frac{1}{4}(n - 2\sqrt{n})$.

If Conjecture 18.9 is true, then the result is sharp as may be seen by considering the following graph. Start with the complete graph on $2k$ vertices, duplicate each edge, and then subdivide each edge. The resultant graph G of order $n = 4k^2$ satisfies $\mathrm{td}(G) = 2\binom{k}{2} = (n - 2\sqrt{n})/4$.

18.9 Dominating and Total Dominating Partitions in Cubic Graphs

Recall that by Theorem 13.4 in Sect. 13.4, every connected cubic graph on n vertices has a TD-set whose complement contains a dominating set such that the cardinality of the TD-set is at most $(n+2)/2$. We conjecture that the constant term can be omitted.

Conjecture 18.10. Every connected cubic graph on n vertices has a total dominating set whose complement contains a dominating set such that the cardinality of the total dominating set is at most $n/2$.

By Proposition 13.12 in Sect. 13.4, if Conjecture 18.10 is true, then the bound is tight. We also conjecture that Conjecture 18.10 is true for graphs with minimum degree at least 3.

Conjecture 18.11. Every connected graph on n vertices with minimum degree at least 3 has a total dominating set whose complement contains a dominating set such that the cardinality of the total dominating set is at most $n/2$.

18.10 Upper Total Domination Versus Upper Domination

By Theorem 16.4 in Chap. 18, if G is a graph with no isolated vertex, then $\Gamma_t(G) \leq 2\Gamma(G)$. Furthermore structural properties of graphs G that achieve equality in this bound are established in Observation 16.5. The following open problem is posed in [127].

Problem 18.1. Find a characterization of graphs G satisfying $\Gamma_t(G) = 2\Gamma(G)$.

18.11 Total Domination Vertex-Critical Graphs

Recall that in Sect. 11.2.5 a graph G is defined as total domination vertex-critical, abbreviated γ_tVC, if for every vertex v of G that is not adjacent to a vertex of degree one we have $\gamma_t(G - v) < \gamma_t(G)$. We pose the following problem.

Problem 18.2. For $k \geq 3$, determine the maximum diameter of a $k_t VC$ graph.

For $k \leq 8$, the value of the maximum diameter of a $k_t VC$ graph is given in Theorem 11.33. However Question 18.3 is still open for all $k \geq 9$. It is shown in Theorem 11.33 that the maximum diameter of a $k_t VC$ is at most $2k-3$, but this bound is most likely far from optimal. We pose the following conjecture.

Conjecture 18.12. The maximum diameter of a $k_t VC$ graph is $\lfloor (5k-7)/3 \rfloor$ for all $k \geq 4$.

Recall that for all $k \equiv 2 \pmod 3$, there exists a $k_t VC$ graph of diameter $(5k-7)/3$, as shown in Theorem 11.34.

18.12 Total Domination Number in Claw-Free Graphs

Recall that in Sect. 9.4 we determine upper bounds on the total domination number of a connected cubic claw-free graph. For an integer $k \geq 2$, let $\mathscr{G}^k_{\text{claw}}$ denote the set of the connected claw-free cubic graphs of order at least $2k$. In [162], Lichiardopol defines

$$f_t(k) = \sup_{G \in \mathscr{G}^k_{\text{claw}}} \frac{\gamma_t(G)}{n(G)}.$$

By Theorem 9.3, we have that $f_t(k) = \frac{1}{2}$ for $k \in \{2,3,4\}$. By Theorem 9.5, $f_t(k) = \frac{4}{9}$ for $k \in \{5,6,7,8,9\}$. By Theorem 9.7, $f_t(10) \leq \frac{10}{23}$. By Theorem 2.11, $\gamma_t(G) \geq n(G)/3$ for every cubic graph G, implying that $f_t(k) \geq \frac{1}{3}$ for all $k \geq 2$. Further, $f_t(k+1) \leq f_t(k)$ for all $k \geq 2$. Hence as remarked by Lichiardopol [162], the sequence $f_t(k)$, $k \geq 2$, converges. Lichiardopol [162] poses the following question.

Question 18.13. What is the value of $\lim_{k \to \infty} f_t(k)$?

18.13 Total Domination Number and Distance

Recall that in Sect. 2.3.4 we discussed bounds on the total domination in terms of the radius and diameter of a graph. We pose the following problem.

Problem 18.3. For each integer $r \geq 3$, determine the largest constant, c_r, such that the following holds for all connected graphs $G = (V,E)$ and for arbitrary $x_1, x_2, \ldots, x_r \in V$:

$$\gamma_t(G) \geq c_r \left(\sum_{1 \leq i < j \leq r} d(x_i, x_j) \right).$$

When $r = 3$ in Question 18.3, $c_3 = 1/4$ by Theorem 2.18. When $r = 4$ in Question 18.3, $c_4 = 1/8$ by Theorem 2.19. However Question 18.3 is still open for all $r \geq 5$. For $r \geq 3$ in general, the following lower and upper bounds on c_r are known.

Theorem 18.14 ([138]). *For $r \geq 3$,*

$$\frac{3}{2r^2 - 2r} \leq c_r \leq \begin{cases} \dfrac{8}{5r^2 - 4r} & \text{if } r \text{ is even} \\ \dfrac{8}{5r^2 - 4r - 1} & \text{if } r \text{ is odd.} \end{cases}$$

18.14 Total Domination Number and Relative Complement

Recall that in Sect. 15.3, we discussed another variation on Nordhaus–Gaddum-type results in which we extended the concept by considering $G_1 \oplus G_2 = K(s,s)$ rather than $G_1 \oplus G_2 = K_n$. In particular, Theorem 15.13 shows that if $G_1 \oplus G_2 = K(s,s)$ where neither G_1 nor G_2 has an isolated vertex, then $\gamma_t(G_1)\gamma_t(G_2) \leq \max\{8s, \lfloor (s+6)^2/4 \rfloor\}$. We pose the following question.

Question 18.15. If $G_1 \oplus G_2 = K(s,s)$ where neither G_1 nor G_2 has an isolated vertex, then is it true that $\gamma_t(G_1)\gamma_t(G_2) \leq 8s$?

If Question 18.15 is true, then the upper bound of $8s$ on the product $\gamma_t^1 \gamma_t^2$ is achievable as may be seen by taking $G_1 = sK_2$ or $G_2 = sK_2$.

18.15 Total Restrained Domination

Recall that a total restrained dominating set, abbreviated TRD-set, in a graph G is a TD-set S in $G = (V,E)$ with the additional property that every vertex not in S is adjacent to a vertex in $V \setminus S$. The minimum cardinality of a TRD-set of G is the total restrained domination number of G, denoted by $\gamma_{tr}(G)$. As shown in Theorem 17.5, if G is a connected cubic graph of order n, then $\gamma_{tr}(G) \leq (n+4)/2$. We pose the following conjecture.

Conjecture 18.16. If G is a connected cubic graph of order n, then $\gamma_{tr}(G) \leq n/2$.

Observation 17.6 shows that if Conjecture 18.16 is true, then the bound is tight.

18.16 Double Total Domination Number

Recall that a double total dominating set, abbreviated DTD-set, of a graph G is defined in Sect. 17.3 as a set S of vertices in G such that every vertex of G is adjacent to at least two vertices in S. The minimum cardinality of a DTD-set in G is the double total domination number $\gamma_{\times 2,t}(G)$ of G. The following conjecture is posed in [135].

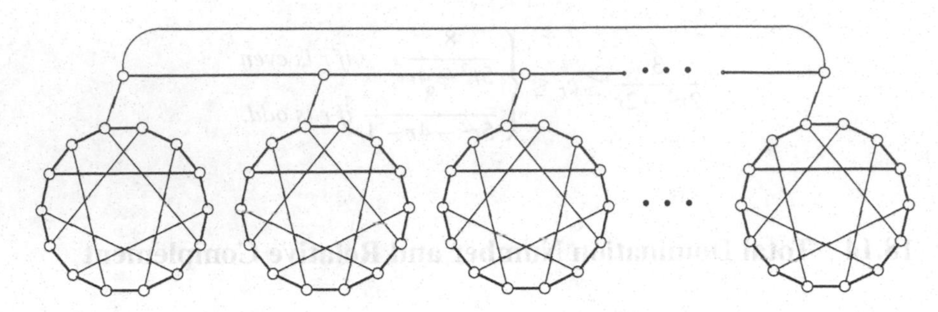

Fig. 18.3 A cubic graph $G \in \mathscr{F}$ of order n with $\gamma_{\times 2,t}(G) = 4n/5$

Conjecture 18.17. If G is a connected graph of sufficiently large-order n with $\delta(G) \geq 3$, then $\gamma_{\times 2,t}(G) \leq 4n/5$.

We remark that if Conjecture 18.17 is true, then the bound is sharp as may be seen by considering the following family $\mathscr{F}$ of all graphs that can be obtained as follows: Take a connected graph F with $\delta(F) \geq 2$ and for each vertex v of F, add a copy of the Heawood graph, and join v to one vertex in that copy of the Heawood graph. A graph G in the family $\mathscr{F}$ is illustrated in Fig. 18.3 (here, F is a cycle). Each graph G of order n in the family $\mathscr{F}$ is a connected graph with $\delta(G) \geq 3$ satisfying $\gamma_{\times 2,t}(G) = 4n/5$.

18.17 Locating–Total Domination

Recall that a locating–total dominating set, abbreviated LTD-set, in a graph $G = (V,E)$ is defined in Sect. 17.4 as a TD-set S with the property that distinct vertices in $V \setminus S$ are totally dominated by distinct subsets of S. The following conjecture was first posed as a question in [123] and subsequently stated as a conjecture in [117].

Conjecture 18.18 ([117]). If G is a connected cubic graph of order $n \geq 8$, then $\gamma_t^L(G) \leq n/2$.

As stated in Sect. 17.4, one might expect that if Conjecture 18.18 is false, then it fails for the class of claw-free graphs since the locating–total domination number of a claw-free cubic graph tends to be very much larger than its total domination number. However as shown in Theorem 17.17, Conjecture 18.18 is true for claw-free graphs. Conjecture 18.18, if true, would be a surprising result since it implies that the tight upper bound on the total domination number of a connected cubic graph of one-half its order is also an upper bound on its locating–total domination number.

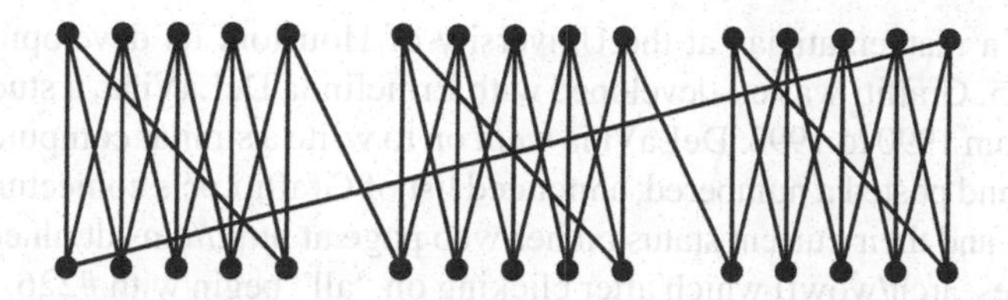

Fig. 18.4 The graph G_3

18.18 Differentiating Total Domination

Recall that in Sect. 17.5 a separating open code in a graph G is a set C of vertices such that $N(u) \cap C \neq N(v) \cap C \neq \emptyset$ for all distinct vertices u and v in G and that an identifying open code, abbreviated IO-code, is a set C that is both a separating open code and a TD-set. Further recall that if G is twin-free and G has no isolated vertex, then $\gamma^{\text{IOC}}(G)$ denotes the minimum cardinality of an IO-code in G. Theorem 17.20 shows that if G is a connected twin-free cubic graph of order n, then $\gamma^{\text{IOC}}(G) \leq 3n/4$. As remarked in Sect. 17.5 this upper bound is achievable, as may be seen by considering the complete graph K_4 or the hypercube Q_3 shown in Fig. 17.4 that satisfies $\gamma^{\text{IOC}}(K_4) = 3$ and $\gamma^{\text{IOC}}(Q_3) = 6$, respectively. However the following conjecture is posed in [140].

Conjecture 18.19 ([140]). If G is a twin-free connected cubic graph of sufficiently large-order n, then $\gamma^{\text{IOC}}(G) \leq 3n/5$.

If Conjecture 18.19 is true, then the upper bound of three-fifths the order is sharp. To see this, an infinite class $\mathscr{G}$ of cubic twin-free connected graphs G satisfying $\gamma^{\text{IOC}}(G) = 3|V(G)|/5$ is constructed in [140]. For $k \geq 1$, let G_k be the graph of order $5k$ constructed as follows. Let

$$V(G) = \{x_0, x_1, \ldots, x_{5k-1}\} \cup \{y_0, y_1, \ldots, y_{5k-1}\}.$$

For every $i = 0, 1, 2, \ldots, k-1$, add the following edges to G_k, where the indices are taken modulo $5k$. If $i \not\equiv 0 \pmod 5$, join x_i to y_{i-1}, y_i and y_{i+1}. If $i \equiv 0 \pmod 5$, join x_i to y_i, y_{i+1} and y_{i+4}. By construction, the graph G_k is a connected cubic graph that is twin-free. The graph G_3, for example, is shown in Fig. 18.4.

It is shown in [140] that for $k \geq 1$, if G_k has order n, then $\gamma^{\text{IOC}}(G_k) = 3n/5$. Therefore if Conjecture 18.19 is true, then the upper bound of three-fifths the order is sharp.

18.19 Graffiti Conjectures

Much interest in total domination in graphs has arisen from a computer program Graffiti.pc that has generated several hundred conjectures on total domination. Graffiti, a computer program that makes conjectures, is a program of Siemion

Fajtlowicz, a mathematician at the University of Houston. Its development began around 1985. Graffiti was co-developed with Ermelinda DeLaViña, a student of Fajtlowicz, from 1990 to 1993. DeLaViña went on to write a similar computer program Graffiti.pc and posted a numbered, annotated list of Graffiti.pc's conjectures on total domination and their current status on her web page at http://cms.dt.uh.edu/faculty/delavinae/research/wowII which after clicking on "all" begin with #226.

References

1. Adhar, G.S., Peng, S.: Parallel algorithms for finding connected, independent and total domination in interval graphs. Algorithms and Parallel VLSI Architectures, II (Château de Bonas, 1991), pp. 85–90. Elsevier, Amsterdam (1992)
2. Alber, J., Bodlaender, H.L., Fernau, F., Niedermeier, R.: Fixed parameter algorithms for planar dominating set and related problems. In: Algorithm theory of Scandinavian Workshop on Algorithm Theory 2000 (Bergen, 2000), pp. 97–110. Springer, Berlin (2000)
3. Alon, N.: Transversal number of uniform hypergraphs. Graphs Combin. **6**, 1–4 (1990)
4. Alon, N., Bregman, Z.: Every 8-uniform 8-regular hypergraph is 2-colorable. Graphs Combin. **4**, 303–306 (1988)
5. Alon, N., Gutner, S.: Linear time algorithms for finding a dominating set of fixed size in degenerated graphs. Algorithmica **54**(4), 544–556 (2009)
6. Aouchiche, M., Hansen, P.: A survey of Nordhaus-Gaddum type relations. Discrete Applied Math. **161**, 466–546 (2013)
7. Archdeacon, D., Ellis-Monaghan, J., Fischer, D., Froncek, D., Lam, P.C.B., Seager, S., Wei, B., Yuster, R.: Some remarks on domination. J. Graph Theory **46**, 207–210 (2004)
8. Arnborg, S., Lagergren, J., Seese, D.: Easy problems for tree-decomposable graphs. J. Algorithms **12**, 308–340 (1991)
9. Bacsó, G.: Complete description of forbidden subgraphs in the structural domination problem. Discrete Math. **309**, 2466–2472 (2009)
10. Bertossi, A.A.: Total domination in interval graphs. Inform. Process. Lett. **23**, 131–134 (1986)
11. Bertossi, A.A., Gori, A.: Total domination and irredundance in weighted interval graphs. SIAM J. Discrete Math. **1**, 317–327 (1988)
12. Bertossi, A.A., Moretti, S.: Parallel algorithms on circular-arc graphs. Inform. Process. Lett. **33**, 275–281 (1990)
13. Blidia, M., Chellali, M., Maffray, F., Moncel, J., Semri, A.: Locating-domination and identifying codes in trees. Australasian J. Combin. **39**, 219–232 (2007)
14. Blidia, M., Dali, W.: A characterization of locating-total domination edge critical graphs. Discussiones Math. Graph Theory **31**(1), 197–202 (2011)
15. Broere, I., Dorfling, M., Goddard, W., Hattingh, J.H., Henning, M.A., Ungerer, E.: Augmenting trees to have two disjoint total dominating sets. Bull. Inst. Combin. Appl. **42**, 12–18 (2004)
16. Bollobás, B., Cockayne, E.J.: Graph-theoretic parameters concerning domination, independence, and irredundance. J. Graph Theory **3**, 241–249 (1979)
17. Brandstädt, A., Kratsch, D.: On domination problems on permutation and other graphs. Theoret. Comput. Sci. **54**, 181–198 (1987)

M.A. Henning and A. Yeo, *Total Domination in Graphs*, Springer Monographs in Mathematics, DOI 10.1007/978-1-4614-6525-6,

18. Brandstädt, A., Le, V.B., Spinrad, J.P.: Graph Classes: A Survey (Monographs in Discrete Mathematics and Applications). SIAM, Philedelphia (1987)
19. Brešar, B., Dorbec, P., Goddard, W., Hartnell, B., Henning, M.A., Klavžar, S., Rall, D.: Vizing's conjecture: A survey and recent results. J. Graph Theory **69**, 46–76 (2012)
20. Brigham, R.C., Carrington, J.R., Vitray, R.P.: Connected graphs with maximum total domination number. J. Combin. Comput. Combin. Math. **34**, 81–96 (2000)
21. Brigham, R.C., Chinn, P.Z., Dutton, R.D.: Vertex domination-critical graphs. Networks **18**, 173–179 (1988)
22. Caccetta, L., Häggkvist, R.: On diameter critical graphs. Discrete Math. **28**(3), 223–229 (1979)
23. Calkin, N., Dankelmann, P.: The domatic number of regular graphs. Ars Combinatoria **73**, 247–255 (2004)
24. Chang, G.J.: Labeling algorithms for domination problems in sun-free chordal graphs. Discrete Appl. Math. **22**, 21–34 (1988)
25. Chang, M.S.: Efficient algorithms for the domination problems on interval and circular-arc graphs. SIAM J. Comput. **27**, 1671–1694 (1998)
26. Chang, M.S., Wu, S.C., Chang, G.J., Yeh, H.G.: Domination in distance-hereditary graphs. Discrete Appl. Math. **116**, 103–113 (2002)
27. Chellali, M.: On locating and differentiating-total domination in trees. Discus. Math. Graph Theory **28**(3), 383–392 (2008)
28. Chellali, M., Favaron, O., Haynes, T.W., Raber, D.: Ratios of some domination parameters in trees. Discrete Math. **308**, 3879–3887 (2008)
29. Chellali, M., Haynes, T.W.: Total and paired-domination numbers of a tree. AKCE Int. J. Graph Comb. **1**, 69–75 (2004)
30. Chellali, M., Haynes, T.W.: A note on the total domination of a tree. J. Combin. Math. Combin. Comput. **58**, 189–193 (2006)
31. Chellali, M., Rad, N.: Locating-total domination critical graphs. Australasian J. Combin. **45**, 227–234 (2009)
32. Chen, X., Liu, J., Meng, J.: Total restrained domination in graphs. Comput. Math. Appl. **62**, 2892–2898 (2011)
33. Chen, X.G., Sohn, M.Y.: Bounds on the locating-total domination number of a tree. Discrete Appl. Math. **159**, 769–773 (2011)
34. Chlebík, M., Chlebíková, J.: Approximation hardness of dominating set problems in bounded degree graphs. Inform. Comput. **206**, 1264–1275 (2008)
35. Chvátal, V., McDiarmid, C.: Small transversals in hypergraphs. Combinatorica **12**, 19–26 (1992)
36. Clark, W.E., Suen, S.: An inequality related to Vizings conjecture. Electronic J. Combin. **7**(Note 4), 3pp (2000)
37. Cockayne, E.J., Dawes, R.M., Hedetniemi, S.T.: Total domination in graphs. Networks **10**, 211–219 (1980)
38. Corneil, D.G., Stewart, L.: Dominating sets in perfect graphs. Discrete Math. **86**, 145–164 (1990)
39. Cockayne, E.J., Dawes, R.M., Hedetniemi, S.T.: Total domination in graphs. Networks **10**, 211–219 (1980)
40. Cockayne, E., Henning, M.A., Mynhardt, C.M.: Vertices contained in all or in no minimum total dominating set of a tree. Discrete Math. **260**, 37–44 (2003)
41. Damaschke, F., Müller, H., Kratsch, D.: Domination in convex and chordal bipartite graphs. Inform. Process. Lett. **36**, 231–236 (1990)
42. Dankelmann, P., Day, D., Hattingh, J.H., Henning, M.A., Markus, L.R., Swart, H.C.: On equality in an upper bound for the restrained and total domination numbers of a graph. Discrete Math. **307**, 2845–2852 (2007)
43. Dankelmann, P., Domke, G.S., Goddard, W., Grobler, P., Hattingh, J.H., Swart, H.C.: Maximum sizes of graphs with given domination parameters. Discrete Math. **281**, 137–148 (2004)

44. DeLaViña, E., Liu, Q., Pepper, R., Waller, B., West, D.B.: Some conjectures of Graffiti.pc on total domination. Congressus Numer. **185**, 81–95 (2007)
45. Desormeaux, W.J., Haynes, T.W., Henning, M.A.: Total domination stable graphs upon edge addition. Discrete Math. **310**, 3446–3454 (2010)
46. Desormeaux, W.J., Haynes, T.W., Henning, M.A.: Total domination critical and stable graphs upon edge removal. Discrete Appl. Math. **158**, 1587–1592 (2010)
47. Desormeaux, W.J., Haynes, T.W., Henning, M.A.: An extremal problem for total domination stable graphs upon edge removal. Discrete Appl. Math. **159**, 1048–1052 (2011)
48. Desormeaux, W.J., Haynes, T.W., Henning, M.A.: Total domination changing and stable graphs upon vertex removal. Discrete Appl. Math. **159**, 1548–1554 (2011)
49. Desormeaux, W.J., Haynes, T.W., Henning, M.A.: Relating the annihilation number and the total domination number of a tree. To appear in Discrete Applied Math.
50. Desormeaux, W.J., Haynes, T.W., Henning, M.A., Yeo, A.: Total domination numbers of graphs with diameter two. To appear in J. Graph Theory
51. Dorbec, P., Gravier, S., Klavžar, S., Špacapan, S.: Some results on total domination in direct products of graphs. Discuss. Math. Graph Theory **26**, 103–112 (2006)
52. Dorbec, P., Henning, M.A., McCoy, J.: Upper total domination versus upper paired-domination. Quaest. Math. **30**, 1–12 (2007)
53. Dorbec, P., Henning, M.A., Rall, D.F.: On the upper total domination number of Cartesian products of graphs. J. Combin. Optim. **16**, 68–80 (2008)
54. Dorfling, M., Goddard, W., Hattingh, J.H., Henning, M.A.: Augmenting a graph of minimum degree 2 to have two disjoint total dominating sets. Discrete Math. **300**, 82–90 (2005)
55. Dorfling, M., Goddard, W., Henning, M.A.: Domination in planar graphs with small diameter II. Ars Combin. **78**, 237–255 (2006)
56. Dorfling, M., Goddard, W., Henning, M.A., Mynhardt, C.M.: Construction of trees and graphs with equal domination parameters. Discrete Math. **306**, 2647–2654 (2006)
57. Dorfling, M., Henning, M.A.: Transversals in 5-uniform hypergraphs and total domination in graphs with minimum degree five. Manuscript
58. Downey, R.G., Fellows, M.R.: Parameterized Complexity. Springer, New York (1999)
59. Duh, R., Fürer, M.: Approximation of *k*-set cover by semi-local optimization. In: Proceedings of the 29th ACM Symposium on Theory of Computing, STOC, pp. 256–264 (1997)
60. El-Zahar, M., Gravier, S., Klobucar, A.: On the total domination number of cross products of graphs. Discrete Math. **308**, 2025–2029 (2008)
61. Fan, G.: On diameter 2-critical graphs. Discrete Math. **67**, 235–240 (1987)
62. Favaron, O., Henning, M.A.: Upper total domination in claw-free graphs. J. Graph Theory **44**, 148–158 (2003)
63. Favaron, O., Henning, M.A.: Paired domination in claw-free cubic graphs. Graphs Combin. **20**, 447–456 (2004)
64. Favaron, O., Henning, M.A.: Total domination in claw-free graphs with minimum degree two. Discrete Math. **308**, 3213–3219 (2008)
65. Favaron, O., Henning, M.A.: Bounds on total domination in claw-free cubic graphs. Discrete Math. **308**, 3491–3507 (2008)
66. Favaron, O., Henning, M.A., Mynhardt, C.M., Puech, J.: Total domination in graphs with minimum degree three. J. Graph Theory **34**(1), 9–19 (2000)
67. Favaron, O., Karami, H., Sheikholeslami, S.M.: Total domination and total domination subdivision number of a graph and its complement. Discrete Math. **308**, 4018–4023 (2008)
68. Fellows, M., Hell, P., Seyffarth, K.: Large planar graphs with given diameter and maximum degree. Discrete Appl. Math. **61**, 133–154 (1995)
69. Feige, U., Halldórsson, M.M., Kortsarz, G., Srinivasan, A.: Approximating the domatic number. SIAM J. Comput. **32**, 172–195 (2002)
70. Füredi, Z.: The maximum number of edges in a minimal graph of diameter 2. J. Graph Theory **16**, 81–98 (1992)
71. Garey, M.R., Johnson, D.S.: Computers and Intractability: A Guide to the Theory of NP-Completeness. W.H. Freeman, San Francisco (1979)

72. Gravier, S.: Total domination number of grid graphs. Discrete Appl. Math. **121**, 119–128 (2002)
73. Goddard, W.: Automated bounds on recursive structures. Util. Math. **75**, 193–210 (2008)
74. Goddard, W., Haynes, T.W., Henning, M.A., van der Merwe, L.C.: The diameter of total domination vertex critical graphs. Discrete Math. **286**, 255–261 (2004)
75. Goddard, W., Henning, M.A.: Domination in planar graphs with small diameter. J. Graph Theory **40**, 1–25 (2002)
76. Goddard, W., Henning, M.A.: Restricted domination parameters in Graphs. J. Combin. Optim. **13**, 353–363 (2007)
77. Goddard, W., Henning, M.A., Swart, H.C.: Some Nordhaus–Gaddum-type results. J. Graph Theory **16**, 221–231 (1992)
78. Hanson, D., Wang, P.: A note on extremal total domination edge critical graphs. Util. Math. **63**, 89–96 (2003)
79. Harary, F.: The maximum connectivity of a graph. Proc. Nat. Acad. Sci. U.S.A. **48**, 1142–1146 (1962)
80. Hartnell, B., Rall, D.F.: Domination in Cartesian products: Vizing's conjecture. In: Haynes, T.W., Hedetniemi, S.T., Slater, P.J. (eds.) Domination in Graphs: Advanced Topics, pp. 163–189. Marcel Dekker, New York (1998)
81. Hattingh, J.H.: Restrained and total restrained domination in graphs. Not. S. Afr. Math. Soc. **41**, 2–15 (2010)
82. Hattingh, J.H., Jonck, E., Joubert, E.J.: An upper bound on the total restrained domination number of a tree. J. Comb. Optim. **20**, 205–223 (2010)
83. Hattingh, J.H., Jonck, E., Joubert, E.J.: Bounds on the total restrained domination number of a graph. Graphs Combin. **26**, 77–93 (2010)
84. Hattingh, J., Joubert, E., van der Merwe, L.C.: The criticality index of total domination of a path. To appear in Util. Math.
85. Haynes, T.W., Hedetniemi, S.T., Slater, P.J.: Fundamentals of Domination in Graphs. Marcel Dekker, New York (1998)
86. Haynes, T.W., Hedetniemi, S.T., Slater, P.J. (eds): Domination in Graphs: Advanced Topics. Marcel Dekker, New York (1998)
87. Haynes, T.W., Henning, M.A.: Trees with unique minimum total dominating sets. Discuss. Math. Graph Theory **22**, 233–246 (2002)
88. Haynes, T.W., Henning, M.A.: Upper bounds on the total domination number. Ars Combinatoria **91**, 243–256 (2009)
89. Haynes, T.W., Henning, M.A.: A characterization of diameter-2-critical graphs with no antihole of length four. Central Eur. J. Math. **10**(3), 1125–1132 (2012)
90. Haynes, T.W., Henning, M.A.: A characterization of diameter-2-critical graphs whose complements are diamond-free. Discrete Applied Math. **160**, 1979–1985 (2012)
91. Haynes, T.W., Henning, M.A.: The Murty Simon conjecture for bull-free graphs, manuscript
92. Haynes, T.W., Henning, M.A.: A characterization of P_5-free, diameter-2-critical graphs, manuscript
93. Haynes, T.W., Henning, M.A., Howard, J.: Locating and total dominating sets in trees. Discrete Appl. Math. **154**, 1293–1300 (2006)
94. Haynes, T.W., Henning, M.A., van der Merwe, L.C., and Yeo, A.: On a conjecture of Murty and Simon on diameter two critical graphs. Discrete Math. **311**, 1918–1924 (2011)
95. Haynes, T.W., Henning, M.A., van der Merwe, L.C., Yeo, A.: On the existence of k-partite or K_p-free total domination edge-critical graphs. Discrete Math. **311**, 1142–1149 (2011)
96. Haynes, T.W., Henning, M.A., van der Merwe, L.C., Yeo, A.: A maximum degree theorem for diameter-2-critical graphs, manuscript
97. Haynes T.W., Henning M.A., van der Merwe L.C., Yeo A.: Progress on the Murty-Simon Conjecture on diameter-2 critical graphs: A Survey. Manuscript.
98. Haynes, T.W., Henning, M.A., Yeo, A.: A proof of a conjecture on diameter two critical graphs whose complements are claw-free. Discrete Optim. **8**, 495–501 (2011)

99. Haynes, T.W., Henning, M.A., Yeo, A.: On a conjecture of Murty and Simon on diameter two critical graphs II. Discrete Math. **312**, 315–323 (2012)
100. Haynes, T.W., Markus, L.R.: Generalized maximum degree. Util. Math. **59**,155–165 (2001)
101. Heggernes, P., Telle, J.A.: Partitioning graphs into generalized dominating sets. Nordic J. Comput. **5**, 128–142 (1998)
102. Henning, M.A.: Graphs with large total domination number. J. Graph Theory **35**(1), 21–45 (2000)
103. Henning, M.A.: Trees with large total domination number. Util. Math. **60**, 99–106 (2001)
104. Henning, M.A.: Restricted domination in graphs. Discrete Math. **254**, 175–189 (2002)
105. Henning, M.A.: Restricted total domination in Graphs. Discrete Math. **289**, 25–44 (2004)
106. Henning, M.A.: A linear Vizing-like relation relating the size and total domination number of a graph. J. Graph Theory **49**, 285–290 (2005)
107. Henning, M.A.: Recent results on total domination in graphs: A survey. Discrete Math. **309**, 32–63 (2009)
108. Henning, M.A.: A short proof of a result on a Vizing-like problem for integer total domination. J. Combin. Optim. **20**, 321–323 (2010)
109. Henning, M.A., Joubert, E.J.: Equality in a linear Vizing-like relation that relates the size and total domination number of a graph, To appear in Discrete Applied Math.
110. Henning, M.A., Joubert, E.J., Southey, J.: Nordhaus–Gaddum bounds for total domination. Appl. Math. Lett. **24**(6), 987–990 (2011)
111. Henning, M.A., Joubert, E.J., Southey, J.: Nordhaus–Gaddum type results for total domination. Discrete Math. Theoret. Comput. Sci. **13**(3), 87–96 (2011)
112. Henning, M.A., Joubert, E.J., Southey, J.: Multiple factor Nordhaus–Gaddum type results for domination and total domination. Discrete Appl. Math. **160**, 1137–1142 (2012)
113. Henning, M.A., Kang, L., Shan, E., Yeo, A.: On matching and total domination in graphs. Discrete Math. **308**, 2313–2318 (2008)
114. Henning, M.A., Kazemi, A.P.: *k*-Tuple total domination in graphs. Discrete Appl. Math. **158**, 1006–1011 (2010)
115. Henning, M.A., Kazemi, A.P.: *k*-Tuple total domination in cross products of graphs. To appear in J. Combin. Optim.
116. Henning, M.A., Löwenstein, C.: Hypergraphs with large transversal number and with edge sizes at least four. Central Eur. J. Math. **10**(3), 1133–1140 (2012)
117. Henning, M.A., Löwenstein, C.: Locating-total domination in claw-free cubic graphs. Discrete Math. **312**,3107–3116 (2012)
118. Henning, M.A., Löwenstein, C., Rautenbach, D.: Partitioning a graph into a dominating set, a total dominating set, and something else. Discussiones Math. Graph Theory **30**(4), 563–574 (2010)
119. Henning, M.A., Löwenstein, C., Rautenbach, D., Southey, J.: Disjoint dominating and total dominating sets in graphs. Discrete Appl. Math. **158**, 1615–1623 (2010)
120. Henning, M.A., Maritz, J.E.: Total restrained domination in graphs with minimum degree two. Discrete Math. **308**, 1909–1920 (2008)
121. Henning, M.A., McCoy, J.: Total domination in planar graphs of diameter two. Discrete Math. **309**, 6181–6189 (2009)
122. Henning, M.A., Rad, N.J.: Total domination vertex critical graphs of high connectivity. Discrete Appl. Math. **157**, 1969–1973 (2009)
123. Henning, M.A., Rad, N.: Locating-total domination in graphs. Discrete Appl. Math. **160**, 1986–1993 (2012)
124. Henning, M.A., Rall, D.F.: On the total domination number of Cartesian products of graph. Graphs Combin. **21**, 63–69 (2005)
125. Henning, M.A., Southey, J.: A note on graphs with disjoint dominating and total dominating sets. Ars Combin. **89**, 159–162 (2008)
126. Henning, M.A., Southey, J.: A characterization of graphs with disjoint dominating and total dominating sets. Quaestiones Math. **32**, 119–129 (2009)

127. Henning, M.A., van der Merwe, L.C.: The maximum diameter of total domination edge-critical graphs. Discrete Math. **312**, 397–404 (2012)
128. Henning, M.A., van der Merwe, L.C.: The maximum diameter of total domination edge-critical graphs. Discrete Math. **312**, 397–404 (2012)
129. Henning, M.A., Yeo, A.: Total domination and matching numbers in claw-free graphs. Electronic J. Combin. **13**, #59 (2006)
130. Henning, M.A., Yeo, A.: A transition from total domination in graphs to transversals in hypergraphs. Quaestiones Math. **30**, 417–436 (2007)
131. Henning, M.A., Yeo, A.: A new upper bound on the total domination number of a graph. Electronic J. Combin. **14**, #R65 (2007)
132. Henning, M.A., Yeo, A.: Total domination in graphs with given girth. Graphs Combin. **24**, 333–348 (2008)
133. Henning, M.A., Yeo, A.: Hypergraphs with large transversal number and with edge sizes at least three. J. Graph Theory **59**, 326–348 (2008)
134. Henning, M.A., Yeo, A.: Total domination in 2-connected graphs and in graphs with no induced 6-cycles. J. Graph Theory **60**, 55–79 (2009)
135. Henning, M.A., Yeo, A.: Strong transversals in hypergraphs and double total domination in graphs. SIAM J. Discrete Math. **24**(4), 1336–1355 (2010)
136. Henning, M.A., Yeo, A.: Perfect matchings in total domination critical graphs. Graphs Combin. **27**, 685–701 (2011)
137. Henning, M.A., Yeo, A.: Girth and total domination in graphs. Graphs Combin. **28**, 199–214 (2012)
138. Henning, M.A., Yeo, A.: A new lower bound for the total domination number n graphs proving a Graffiti Conjecture. Manuscript
139. Henning, M.A., Yeo, A.: Total domination and matching numbers in graphs with all vertices in triangles. Discrete Math. **313**, 174–181 (2013)
140. Henning, M.A., Yeo, A.: Identifying open codes in cubic graphs. To appear in Graphs Combin.
141. Henning, M.A., Yeo, A.: 2-Colorings in *k*-regular *k*-uniform hypergraphs. To appear in European J. Combin.
142. Ho, P.T.: A note on the total domination number. Util. Math. **77**, 97–100 (2008)
143. Hoffman, A.J., Singleton, R.R.: On Moore graphs with diameter 2 and 3. IBM J. Res. Develop. **5**, 497–504 (1960)
144. Honkala, I., Laihonen, T., Ranto, S.: On strongly identifying codes. Discrete Math. **254**, 191–205 (2002)
145. Imrich, W., Klavžar, S.: Product Graphs: Structure and Recognition. Wiley, New York (2000)
146. Jiang, H., Kang, L.: Total restrained domination in claw-free graphs. J. Comb. Optim. **19**, 60–68 (2010)
147. Jiang, H., Kang, L., Shan, E.: Total restrained domination in cubic graphs. Graphs Combin. **25**, 341–350 (2009)
148. Jiang, H., Kang, L., Shan, E.: Graphs with large total restrained domination number. Util. Math. **81**, 53–63 (2010)
149. Joubert, E.J.: Total restrained domination in claw-free graphs with minimum degree at least two. Discrete Appl. Math. **159**, 2078–2097 (2011)
150. Keil, J.M.: Total domination in interval graphs. Inform. Process. Lett. **22**, 171–174 (1986)
151. Keil, J.M.: The complexity of domination problems in circle graphs. Discrete Appl. Math. **42**, 51–63 (1993)
152. Khodkar, A., Mojdeh, D.A., Kazemi, A.P.: Domination in Harary graphs. Bull. ICA **49**, 61–78 (2007)
153. König, D.: Über Graphen und ihre Anwendung auf Determinantheorie und Mengenlehre. Math. Ann. **77**, 453–465 (1916)
154. Kratsch, D., Stewart, L.: Domination on cocomparability graphs. SIAM J. Discrete Math. **6**, 400–417 (1993)

155. Kratsch, D., Stewart, L.: Total domination and transformation. Inform. Process. Lett. **63**, 167–170 (1997)
156. Kratsch, D.: Domination and total domination on asteroidal triple-free graphs. Proceedings of the 5th Twente Workshop on Graphs and Combinatorial Optimization (Enschede, 1997). Discrete Appl. Math. **99**, 111–123 (2000)
157. Lam, P.C.B., Wei, B.: On the total domination number of graphs. Util. Math. **72**, 223–240 (2007)
158. Laskar, R.C., Pfaff, J.: Domination and irredundance in split graphs. Technicial Report 430, Clemson University, Dept. Math. Sciences (1983)
159. Laskar, R.C., Pfaff, J., Hedetniemi, S.M., Hedetniemi, S.T.: On the algorithmic complexity of total domination. SIAM J. Algebr. Discrete Meth. **5**, 420–425 (1984)
160. Li, N., Hou, X.: On the total $\{k\}$-domination number of Cartesian products of graphs. J. Comb. Optim. **18**, 173–178 (2009)
161. Li, N., Hou, X.: Total domination in the Cartesian product of a graph and K_2 or C_n. Util. Math. **83**, 313–322 (2010)
162. Lichiardopol, N.: On a conjecture on total domination in claw-free cubic graphs: proof and new upper bound. Australasian J. Combin. **51**, 7–28 (2011)
163. Loizeaux, M., van der Merwe, L.: A total domination vertex-critical graph of diameter two. Bull. ICA **48**, 63–65 (2006)
164. MacGillivray, G., Seyffarth, K.: Domination numbers of planar graphs. J. Graph Theory **22**, 213–229 (1996)
165. McRae, A.A.: Generalizing NP-completeness proofs for bipartite and chordal graphs. PhD thesis, Clemson University (1994)
166. Meir, A., Moon, J.W.: Relations between packing and covering numbers of a tree. Pacific J. Math. **61**, 225–233 (1975)
167. Mitchell, S.L., Cockayne, E.J., Hedetniemi, S.T.: Linear algorithms on recursive representations of trees. J. Comput. Syst. Sci. **18**, 76–85 (1979)
168. Nordhaus, E.A., Gaddum, J.W.: On complementary graphs. Amer. Math. Monthly **63**, 175–177 (1956)
169. Nowakowski, R.J., Rall, D.F.: Associative graph products and their independence, domination and coloring numbers. Discuss. Math. Graph Theory **16**, 53–79 (1996)
170. Pfaff, J., Laskar, R.C., Hedetniemi, S.T.: NP-completeness of total and connected domination and irredundance for bipartite graphs. Technicial Report 428, Dept. Math. Sciences, Clemson University (1983)
171. Philip, G.: The kernelization complexity of some domination and covering problems. Ph.D. Thesis
172. Plesník, J.: Critical graphs of given diameter. Acta F.R.N Univ. Comen. Math. **30**, 71–93 (1975)
173. Plesní, k.P.: Bounds on chromatic numbers of multiple factors of a complete graph. J. Graph Theory **2**, 9–17 (1978)
174. Pradhan, D.: Complexity of certain functional variants of total domination in chordal bipartite graphs. Discrete Math. Algorithms Appl. **4**(3), 1250045 (19 p) (2012)
175. Rall, D.F.: Total domination in categorical products of graphs. Discuss. Math. Graph Theory **25**, 35–44 (2005)
176. Ramalingam, G., Rangan, C.P.: Total domination in interval graphs revisited. Inform. Process. Lett. **27**, 17–21 (1988)
177. Rao, A., Rangan, C.P.: Optimal parallel algorithms on circular-arc graphs. Inform. Process. Lett. **33**, 147–156 (1989)
178. Robertson, N., Seymour, P.D.: Graph minors. II. Algorithmic aspects of tree-width. J. Algorithms **7**(3), 309–322 (1986)
179. Sanchis, L.A.: Bounds related to domination in graphs with minimum degree two. J. Graph Theory **25**, 139–152 (1997)
180. Schaudt, O.: On the existence of total dominating subgraphs with a prescribed additive hereditary property. Discrete Math. **311**, 2095–2101 (2011)

181. Sanchis, L.A.: Relating the size of a connected graph to its total and restricted domination numbers. Discrete Math. **283**, 205–216 (2004)
182. Seo, S.J., Slater, P.J.: Open neighborhood locating-dominating sets. Australasian J. Combin. **46**, 109–120 (2010)
183. Seo, S.J., Slater, P.J.: Open neighborhood locating-dominating in trees. Discrete Appl. Math. **159**, 484–489 (2011)
184. Shan, E., Kang, L., Henning, M.A.: Erratum to: A linear Vizing-like relation relating the size and total domination number of a graph. J. Graph Theory **54**, 350–353 (2007)
185. Singleton, R.R.: There is no irregular Moore graph. Amer. Math. Monthly **75**, 42–43 (1968)
186. Slater, P.J.: Dominating and location in acyclic graphs. Networks **17**, 55–64 (1987)
187. Slater, P.J.: Dominating and reference sets in graphs. J. Math. Phys. Sci. **22**, 445–455 (1988)
188. Southey, J., Henning, M.A.: On a conjecture on total domination in claw-free cubic graphs. Discrete Math. **310**, 2984–2999 (2010)
189. Southey, J., Henning, M.A.: Dominating and total dominating partitions in cubic graphs. Central Eur. J. Math. **9**(3), 699–708 (2011)
190. Southey, J., Henning, M.A.: An improved upper bound on the total restrained domination number in cubic graphs. Graphs Combin. **8**, 547–554 (2012)
191. Southey, J., Henning, M.A.: Edge weighting functions on dominating sets. To appear in J. Graph Theory
192. O, S., West, D.B.: Balloons, cut-edges, matchings, and total domination in regular graphs of odd degree. J. Graph Theory **64**, 116–131 (2010)
193. Sumner, D.P., Blitch, P.: Domination critical graphs. J. Combin. Theory Ser. B **34**, 65–76 (1983)
194. Sun, L.: An upper bound for the total domination number. J. Beijing Inst. Tech. **4**, 111–114 (1995)
195. Telle, J.A.: Complexity of domination-type problems in graphs. Nordic J. Comput. **1**, 157–171 (1994)
196. Telle, J.A., Proskurowski, A.: Algorithms for vertex partitioning problems on partial *k*-trees. SIAM J. Discrete Math. **10**, 529–550 (1997)
197. Thomassé, S., Yeo, A.: Total domination of graphs and small transversals of hypergraphs. Combinatorica **27**, 473–487 (2007)
198. Thomassen, C.: The even cycle problem for directed graphs. J. Am. Math. Soc. **5**, 217–229 (1992)
199. Turán, P.: Eine Extremalaufgabe aus der Graphentheorie. Mat. Fiz. Lapok **48**, 436–452 (1941)
200. Tuza, Z.s.: Covering all cliques of a graph. Discrete Math. **86**, 117–126 (1990)
201. Tuza, Z.s.: Hereditary domination in graphs: characterization with forbidden induced subgraphs. SIAM J. Discrete Math. **22**, 849–853 (2008)
202. Van der Merwe, L.C.: Total domination edge critical graphs. Ph.D. Dissertation, University of South Africa (1998)
203. Van der Merwe, L.C., Haynes, T.W., Mynhardt, C.M.: Total domination edge critical graphs. Util. Math. **54**, 229–240 (1998)
204. Van der Merwe, L.C., Loizeaux, M.: 4_t-Critical graphs with maximum diameter. J. Combin. Math. Combin. Comput. **60**, 65–80 (2007)
205. Wang, H., Kang, L., Shan, E.: Matching properties on total domination vertex critical graphs. Graphs Combin. **25**, 851–861 (2009)
206. Xu, J.: A proof of a conjecture of Simon and Murty (in Chinese). J. Math. Res. Exposition **4**, 85–86 (1984)
207. Yeo, A.: Relationships between total domination, order, size and maximum degree of graphs. J. Graph Theory **55**(4), 325–337 (2007)
208. Yeo, A.: Manuscript
209. Yuster, R.: The domatic number of regular and almost regular graphs, mansucript (see http://arxiv.org/pdf/math/0111257v1.pdf)
210. Vizing, V.G.: A bound on the external stability number of a graph. Dokl. Akad. Nauk SSSR **164**, 729–731 (1965)

211. Zelinka, B.: Total domatic number and degrees of vertices of a graph. Math. Slovaca **39**, 7–11 (1989)
212. Zelinka, B.: Domatic numbers of graphs and their variants: A survey. In: Haynes, T.W., Hedetniemi, S.T., Slater, P.J. (eds.) Domination in Graphs: Advanced Topics, pp. 351–377. Marcel Dekker, New York (1998)

Glossary

$\square$:	The Cartesian product. 96, 104, 106, 107, 137, 139
$\boxdot$:	See page 85 for a definition. 85–88
$\gamma\gamma_t(G)$:	Defined as $\min\{\vert D\vert + \vert T\vert : (D,T)$ is DT-pair of $G\}$. 111, 112
$\Delta(G)$:	Maximum degree in the graph G. 3
$\Delta(H)$:	Maximum degree in the hypergraph H. 6, 12
$\Delta_k(G)$:	Generalized maximum degree in G. 15
$\Gamma_t(G)$:	Upper total domination number. 2
$\Gamma_t(G)$-set:	A minimal TD-set of cardinality $\Gamma_t(G)$. 2
$H \circ K_1$:	The corona, $\mathrm{cor}(H)$, of H. 5
$H \circ P_2$:	The 2-corona of H. 5
$\alpha_2(G)$:	The size of a maximum set inducing a graph with maximum degree at most one. 131
$\alpha(G)$:	The independence number of G. 131, 132
$\alpha'(G)$:	The matching number of G. 5, 77–81
$\overline{G}$:	The complement of a graph. 3, 83, 84, 89, 92, 93, 95, 97, 125–128
$\delta_1(G;X,Y)$:	The number of degree-1 vertices in G that do not belong to Y. 60, 61
$\delta_{2,1}(G;X,Y)$:	The number of degree-2 vertices in G that do not belong to $X \cup Y$ and are adjacent to a degree-1 vertex in G that also does not belong to $X \cup Y$. 60, 61
$\delta(G)$:	Minimum degree of a vertex in the graph G. 3

M.A. Henning and A. Yeo, *Total Domination in Graphs*, Springer Monographs in Mathematics, DOI 10.1007/978-1-4614-6525-6,

$\delta^*(G)$: .... Defined as $\min\{\delta(G), \delta(\overline{G})\}$. 125–128

$\delta(H)$: .... Minimum degree in the hypergraph H. 6

dom(G): .... Domatic number. 109, 110

$\gamma(G)$: .... Domination number. 2, 17, 36, 55, 56, 77, 96, 98, 106

γVC: .... γ-vertex-critical. 96, 97

$\gamma\gamma_t^*$: .... The DT-pair total domination number. 115, 116

$\gamma\gamma_t^*(G)$-set: .... A DT-pair total dominating set of cardinality $\gamma\gamma_t^*(G)$. 115

$\gamma(G)$-set: .... A dominating set of cardinality $\gamma(G)$. 2, 17, 56

γ_tEC: .... Total domination edge critical. 87

$\gamma_t(G)$: .... Total domination number. 1

$\gamma_t(G;X)$: .... The almost total domination number of G relative to X. 85

$\gamma_t^*(G)$: .... Defined as $\min\{\gamma_t(G), \gamma_t(\overline{G})\}$. 125, 126

γ_tVC: .... Total domination vertex critical. 95–97, 153

$\gamma_t(G;X,Y)$: .... (X,Y)-total domination number. 59, 61

$\gamma_t(G;X,Y)$-set: ...... A (X,Y)-total dominating set of size $\gamma_t(G;X,Y)$. 59

γ_t^--stable: .... Total domination edge removal stable. 99, 100

γ_t^--vertex changing: .. Total domination vertex removal changing graphs. 100, 101

γ_t^--vertex stable: ..... Total domination vertex removal stable graphs. 101

γ_t^+-stable: .... total domination edge addition stable. 97–99

$\gamma_t(G)$-set: .... A TD-set of cardinality $\gamma_t(G)$. 1

γ_t-special: .... There exist two distinguished vertices u and v in G such that $N[u] = N[v]$ and $\gamma_t(G;x) = \gamma_t(G)$ for $x \in \{u,v\}$. 85, 86

γ_t-vertex-critical: ..... a vertex v is γ_t-vertex-critical if $\gamma_t(G-v) < \gamma_t(G)$. 95

γ-vertex-critical: ..... k-γ-critical for some k. 96

k_t^--stable: .... Total domination number k and γ_t^--stable. 99

$\rho(G)$: .... The packing number of G. 4, 36

(ρ, γ_t)-graph: .... See page 36 for a definition. 36

(ρ, γ_t)-labeling: ..... See page 36 for a definition. 36

$\rho^o(G)$: .... The open packing number of G. 4, 33, 107

$\succ$: .... Dominates. 2, 94

$\succ_t$: .... Totally dominates. 1

τ: .... transversal number. 7, 8, 12, 13, 39, 43, 44, 48, 51, 52, 122, 123, 151, 152

$\mathrm{td}(G)$: .... The number of edges that must be added to G to ensure a partition of $V(G)$ into two TD-sets. 110, 111

$\mathrm{tdom}(G)$: .... Total domatic number. 109, 110

$\mathscr{A}_t(G)$: .... All vertices that belong to every $\gamma_t(G)$-set in G. 33, 101

$\mathrm{bc}(G)$: .... $\mathrm{bc}(G;X,Y)$ when $X = Y = \emptyset$. 60, 61

$\mathrm{bc}(G;X,Y)$: .... The number of (X,Y)-cut-vertices in G. 60, 61

$B(G)$: .... The periphery of G. 16

C_4-free: .... Contains no 4-cycle as an induced subgraph. 91, 92, 150

C_5-free: .... Contains no 5-cycle as an induced subgraph. 112

C_6-free: .... Contains no 6-cycle as an induced subgraph. 59, 61

C_k-free: .... Contains no k-cycle as an induced subgraph. 92

C_n: .... A cycle of length n. 4, 14, 106, 108

$C(v)$: .... The children of v in a tree. 5

$C(G)$: .... The center of G. 16

$\mathrm{cor}(H)$: .... The corona of H. 5, 95, 111

$D(v)$: .... The descendants of v in a tree. 5, 34

$D[v]$: .... Defined as $\{v\} \cup D(v)$. 5

$\mathrm{diam}(G)$: .... The diameter of G. 4, 16, 55, 56, 58, 87, 90, 98

$\gamma_{\times 2,t}(G)$: .... The double total domination number. 144, 155, 156

$e_i(H)$: .... The number of i-edges in H. 50

$\mathrm{ecc}(B(G))$: .... The eccentricity of the periphery in G. 16

$\mathrm{ecc}(C(G))$: .... The eccentricity of the center in G. 16

$\mathrm{ecc}_G(S)$: .... The eccentricity of a set S of vertices in G. 4

$\mathrm{ecc}_G(v)$: .... The eccentricity of a vertex v in G. 4

epn(v,S): .... The S-external private neighborhood of v (also denoted by epn$[v,S]$). 3, 9, 10, 12, 98, 136, 139

FPT: .... Fixed parameter tractability. 21–23

G_{10}: .... The Petersen graph. 2, 115, 116

G_{16}: .... The generalized Petersen graph. 6–8, 44, 67, 72, 116, 150

G^h: .... The G-fiber (given vertex $h \in V(H)$). 104, 139

${}^gH$: .... The H-fiber (given vertex $g \in V(G)$). 104, 105, 139

$G \square H$: .... The Cartesian product of G and H. 4, 103

$G \oplus H$: .... Is defined as $V(G \oplus H) = V(G) = V(H)$ and $E(H) = E(G) \cup E(H)$. 4, 128

$G \otimes H$: .... Graph product of G and H. 103, 107

$G \times H$: .... The direct product of G and H. 107

$\gamma_t^L(G)$: .... Locating-total domination number. 31, 145

ipn$[v,S]$: .... The S-internal private neighborhood of v. 3

ipn(v,S): .... The open S-internal private neighborhood of v. 3, 9, 10, 98, 136

$K_{1,3}$-free: .... Claw-free. 4

$K_{1,5}$-free: .... Contains no $K_{1,5}$ as an induced subgraph. 95, 96

$K_{1,6}$-free: .... Contains no $K_{1,6}$ as an induced subgraph. 96

$K_{1,7}$-free: .... Contains no $K_{1,7}$ as an induced subgraph. 96

$K_{1,8}$-free: .... Contains no $K_{1,8}$ as an induced subgraph. 96

K_3-free: .... Triangle free (also C_3-free). 94

k-γ-critical: .... Vertex domination critical with $\gamma = k$. 96

$K_{n,m}$: .... A complete bipartite graph. 4

K_p-free: .... Contains no complete graph of order p. 94

k_tEC: .... Total domination edge critical and $\gamma_t(G) = k$. 83, 87, 88, 149

k_t^+-stable: .... Total domination number k and γ_t^+-stable. 97, 98

k_tVC: .... Total domination vertex-critical and $\gamma_t(G) = k$. 95, 153, 154

LTD-set: .... Locating-total dominating set. 31, 32

$N_G(v)$: .... The open neighborhood of v. 1

$N_G[v]$: .... The closed neighborhood of v. 1

$N(S)$: .... The open neighborhood of the set S. 1

$N[S]$: .... The closed neighborhood of the set S. 1

$\mathscr{N}_t(G)$: .... All vertices that belong to no $\gamma_t(G)$-set in G. 33

$OB(S)$: .... Open boundary of S. 10

$\mathrm{oc}(G)$: .... The number of odd components in G. 5

ONH: .... Open neighborhood hypergraph. 2, 6–8, 13, 39, 44, 48, 51, 52

P_5-free: .... Contains no P_5 as an induced subgraph. 93

P_n: .... A path on n vertices. 4, 14, 106, 107, 131

$\mathrm{pc}(G)$: .... The path covering number of G. 78

$\mathrm{pn}[v,S]$: .... The S-private neighborhood of v. 3

$\mathrm{pn}(v,S)$: .... The open S-private neighborhood of v. 3, 9, 35

$\mathrm{rad}(G)$: .... The radius of G. 4, 15, 16, 58

$S(G)$: .... The set of support vertices in G. 100

TD-set: .... Total dominating set. 1, 9

$Total(\mathscr{G})$: .... See page 10 for a definition. 10, 11

(u,v)-path: .... A path from u to v. 4

$W[2]$-hard: .... A hard problem for complexity class $W[2]$. 22

$[X,Y]$: .... Edges between X and Y. 4

(X,Y)-cut-vertex: .... See page 60 for a definition. 60

(X,Y)-TD-set: .... (X,Y)-total dominating set. 59, 60

(X,Y)-total dominating set: A set of vertices S such that $X \cup Y \subseteq S$ and S totally dominates all vertices not in Y. 59

(X,Y)-total domination number: The minimum size of an (X,Y)-total dominating set. 59

Index

M.A. Henning and A. Yeo, *Total Domination in Graphs*, Springer Monographs in Mathematics, DOI 10.1007/978-1-4614-6525-6,

Printed in the United States
By Bookmasters